Hortus Eystettensis

Hortus Eystettensis
The Bishop's Garden
AND BESLER'S MAGNIFICENT BOOK

NICOLAS BARKER

THE BRITISH LIBRARY

First published 1994
by The British Library
Great Russell Street, London WC1B 3DG

Published in North and South America
by Harry N. Abrams Inc., New York

© 1994 Nicolas Barker

Cataloguing in Publication Data
is available from The British Library

ISBN 0 7123 0339 1

Designed by John Mitchell
Typeset in Linotype Caslon
by Bexhill Phototypesetters, Bexhill-on-Sea
Printed in Italy by Artegrafica, Verona

CONTENTS

Foreword *page* vi

PROLOGUE
The Garden of Eichstatt: an eye-witness's account *page* 1

THE HISTORY OF THE BOOK
The Bishop and his garden *page* 5
Printing and publication: the evidence of the book *page* 9
Printing and publication: the archival sources *page* 14
Later history *page* 18

CREATING THE PLATES
The sources of the plates *page* 23
Transfer from nature to plate *page* 28

COLOURING THE BOOK
The early coloured copies *page* 35
The later coloured copies *page* 43

CATALOGUE OF THE COLOURED COPIES
page 49

APPENDICES
A: Documents *page* 60
B: Record of colour variants *page* 68
C: Engravers' and illuminators' signatures *page* 73
D: Analysis of the Schedel Kalendarium *page* 76
E: Colour reversal *page* 79
F: Chemical analysis of the pigments used in modern copies *page* 79

Notes on the plates *page* 80

THE PLATES

Foreword

HORTUS EYSTETTENSIS was the name of a garden that grew round a castle on top of a rock, the palace of Johann Conrad von Gemmingen, Bishop of Eichstätt. It was encircled by a river that separated it from the town of Eichstätt, half way between Nürnberg and Munich, fifty miles away to the north and south, respectively. It was also the name of a great book, the largest ever, at the beginning of the seventeenth century, devoted to the depiction of plants. The pictures were black and white, a masterpiece of the delicate art of engraving on copper. Flowers and plants might wither and die, but the leaves of the book preserved their images, vividly recalling them to those fortunate enough to possess it. A few of these, even more fortunate, had the pictures painted in life-like colours, by artists as skilled as the engravers; between twenty and thirty of these still survive, in whole or part, greatest of all monuments of horticulture and botany as they were then conceived. In all this we see the hand of an impresario of courage and imagination, Basilius Besler, apothecary of Nürnberg.

How all these books came into existence makes a complicated story, which begins when the medieval art of illumination was still alive and ends not long before the invention of photography. Much of the evidence for it lies in the copies themselves, easy to see but hard to recognize. Piecing it together is a new form of textual criticism. The texts on which critics conventionally work consist of words, and their task is to trace the filiation of the different forms in which the words are found. The lines of descent or ascent thus achieved have, like a good family tree, the reassuring conviction, the finality, of a mathematical equation. If the primal form, the archetype, can be discovered, or even its now lost form deduced, the edifice looks as solid as the pyramid it resembles.

In fact, however, matters are seldom so simple or reliable. Are these errors the cumulative result of many stages of degradation, or are they an honest but incompetent record of a very early state of the text in a hand now so unfamiliar as to be almost illegible? Is that immaculate version the author's final text, or an exercise in sophisticated rationalization by a later hand? These uncertainties are common to the editors of verbal texts; they are even more familiar to those who deal in the same way with visual images. Which comes first, the drawing or the painting? Does a rough unfinished look indicate the artist's first thoughts or an inept copy? If two pictures are reverse images of each other, which is the earlier, which the later? All these questions, and others besides, occurred and recurred in the course of surveying the coloured copies of the *Hortus Eystettensis*.

The novelty, in this case, lay in the fact that the images, fixed by the engraver's burin, do not change, only the colours and the way they are applied. These changes may seem what the textual critic would call 'accidentals', as opposed to the substantive change that altered images would imply. Differences in accidentals are always hard to interpret. Are they indeed accidental, an unconscious change? Or are they deliberate, a conscious correction? Or, again, are they individual tokens of some larger scheme, the application of a new system or convention to convey the sense of the text, whether verbal or visual? Accidents did indeed take place. The change from copying a coloured model to following written instructions caused characteristic errors. Other individual changes were made in response to a new sense of naturalism. There was a more pervasive change from a strong almost three-dimensional sense of colour, aimed at capturing the thing itself on paper, to a more delicate *aquarelliste* style in which paint and paper conspire to cheat the eye with an illusion, not reality. It is not easy to identify signs like these, still less to interpret them. But from them the sequence of the coloured copies can be established, and, once established, gives point to a whole range of facts, drawn from other sources as well, about the *Hortus Eystettensis* over two centuries.

No one, embarking on such a study, can begin without the acknowledgement due to the pioneers, Christoph Jakob Trew in the eighteenth century and Joseph Schwertschlager in the nineteenth, but this quest began with Hans Baier's short but pregnant article in *Aus dem Antiquarlat* in 1970, 'Die Ausgaben des Hortus Eystettensis 1613–1750'. This was the first attempt to describe its several appearances, to relate them to contemporary records, and to provide a census of surviving copies. It was a notable achievement, the more so as it was largely conducted by correspondence. If I have come to doubt its conclusions on the first two topics, I can never be sufficiently grateful for the census which led the way to many of the copies described here.

These descriptions have only been possible thanks to the existence of a facsimile. *The Besler Florilegium*, the French *Herbier de quatre saisons,* is a complete reproduction of the copy of *Hortus Eystettensis* in the Muséum de Histoire Naturelle in Paris. It was not ideal, for many of its colour 'readings' are peculiar to it, but without it the essential task of comparing the different colouring of 367 plates in over twenty copies would have been impossible. My copy of the facsimile is now covered with notes in many different coloured inks, recording these differences. I have used its consecutive renumbering, instead of the complex original system of seasons and *ordines,* to refer to individual plates, silently correcting the transposition of two pairs of plates, 92 and 131, and 322 and 323, misplaced in the original.

The layers of information thus constructed are the result of many journeys with my wife who shared the task of turning the huge pages and noting the details, during which we incurred many debts of help and hospitality which I can record if not repay. The first began at Wiesbaden where Dr W. Podehl welcomed us to the Hessische Landesbibliothek. After Wiesbaden we went to Mainz, where I saw the only coloured copy of the 1640 edition, and thence to Speyer, where Dr H. Harthausen showed us the 1750 copy in the Pfalzische Landesbibliothek. With this preamble we came finally to Nürnberg, the origin of the book, where, then and since, Elisabeth Beare made us welcome at the Stadtbibliothek and elsewhere. Her generous support has lightened the task immeasurably, and her sympathetic insight led to new prospects, a larger and better view of our subject.

From Nürnberg a suburban train runs to Erlangen, whose University Library is the centre of Beslerian studies. Trew's collections, the drawings made for the plates and many other documents, have been added to the famous Altdorf copy of the *Hortus*, not to mention the earlier drawings of Oelinger and Gesner. Recently, thanks to the KulturStiftung der Länder, they have been joined by the 'Florilegium' associated with Joachim Camerarius the younger, sold at Christie's on 22 May 1992. Through all these riches Dr Hans-Otto Keunecke has been a sympathetic and informative guide; in 1989 he and Dr Konrad Wickert put on the admirable exhibition, *Hortus Eystettensis*, whose catalogue has been an invaluable source. West of Nürnberg lie Ansbach and Ellwangen, where Frau Schäfer made us welcome, and then the way led south to Eichstätt itself, where Armin Jedlitschka greeted us at the University Library, and we were able to see the site of the garden (the castle was extensively rebuilt in the nineteenth century). Then came Munich, where Dr Hertrich showed us the copy there, and finally Berlin, where alone there are two copies, and the double task was made light by the kindness and help of Friedhilde Krause and Renate Schipke.

Outside Germany, my friends at the Bibliothèque Nationale in Paris, particularly Jean Toulet and Ursula Baurmeister, have been continuously helpful, as has the staff of the Muséum de Histoire Naturelle. In Vienna, Otto Mazal and Ernst Gamillschegg answered many questions. George Colin and Elly Cockx-Indestege welcomed us at the Bibliothèque Royale, Brussels, as did Dr R. Breugelmans at Leiden University. The staff of the Biblioteca Nacional, Madrid, in particular Adela Paz, were endlessly kind. Heribert Tenschert was a hospitable host at the Bibermuhle, and W. Graham Arader III kindly showed me the surviving leaves of the copy sold at Monte Carlo in 1986. Fr Leonard Boyle, with characteristic generosity, twice allowed me to examine the Vatican copy in his own room. I have not, as yet, seen the copy at Leufsta, but Tomas Anfält at the Uppsala University Library became my second self, recording details and supplying photographs with equal enthusiasm. Many others, too many to record here, have helped my quest, and I hope they will accept a general acknowledgement.

I owe more than I can say to the two private owners, who provided hospitality and photographs with equal liberality; without their aid this work would have been vastly poorer. The Oak Spring Library is an institution in itself, but my personal gratitude to Mr and Mrs Paul Mellon for their kindness and hospitality is great. I have a special debt also to three friends, who together made it possible to recover Sebastian Schedel's part in the making of the *Hortus Eystettensis*. Heidrun Ludwig at Nürnberg shared the information gathered for her own work on German botanical illustration, not only about Schedel but Magdalena Fürstin and Hans Thomas Fischer. Sue Abbe Kaplan allowed me to use her thesis on Schedel's masterpiece, the 'Schembartlauf' manuscript in the University Reference Library, University of California, Los Angeles. Cheryl Piggott, the archivist at Kew, gave me generous access to his 'Kalendarium', and thus to the origins of the work. A chance visit to Bamberg with Elisabeth Beare revealed the parallel enterprise of Abbot Johann Müller; there, Bernhard Schemmel, director of the Staatsbibliothek, and Werner Dressendorfer, apothecary, gave generously both time and photographs. Cheryl Porter kindly provided access to the Chemistry Department's facilities at University College London, and the report on the example of modern colouring. Anne Simon undertook the translation of the documents; I should have been lost without her skill in deciphering texts difficult to understand even when translated.

Sandra Raphael, who kindly read the proofs, and Rick Watson gave all manner of help and advice at every stage of my work, to its great benefit. Kathleen Houghton, John Mitchell, whose design skill I have long admired, and David Way, at the British Library, Paul Gottlieb and Marty Malovany of Harry Abrams, all made the process of publication a pleasure. But my last and greatest debt is to the firm of Bernard Quaritch Ltd, and all its staff, in particular the chairman, Lord Parmoor, Nicholas Poole-Wilson, managing director, Wendy Delamore and Detlef Auvermann. What began as a modest attempt to examine and if possible vindicate their judgement of the De Belder copy which they sold in 1987 has grown to its present form. Their patience and support never waned and I hope that this book will now justify their faith in it.

I know, however, that it will not be the last word on the *Hortus Eystettensis*. I am uneasily aware that at least one other coloured copy exists that I have not been able to see; more drawings besides Schedel's may anonymously wait discovery in public or private collections. The tenuous links that seem to join them to the early coloured copies and those to others painted a century and a half later may be disproved or strengthened. I hope, at least, that some of the evidence collected here may enable further research into the greatest botanical book of its time.

NICOLAS BARKER

PROLOGUE

The Garden of Eichstätt: an eye-witness's account

ON 17 May 1611 Philipp Hainhofer saw the gardens of the Prince Bishop of Eichstätt, Johann Conrad von Gemmingen, and recorded what he saw in the travel diary that he kept. He was then 32 years old, and the Bishop 50 or 51, but already mortally ill. The circumstances of this visit were complicated. Hainhofer came of a merchant family in Augsburg. His student years were spent in Italy, but he had also travelled in France and the Low Countries; he spoke all three languages, but Italian most easily (it breaks out in his journal, and he used it to correspond with his more educated patrons). He had acquired a taste for works of art of all sorts, and on his return to Augsburg set up a business, collecting such things himself and acting as an agent for their collection by kings, princes, noblemen and ambassadors. Like many another well-travelled man of his time he was a writer of newsletters, the semi-official agent of Wilhelm V, Duke of Bavaria, Henry IV of France, the Markgraf of Baden, and, in particular the Dukes Philipp II of Pomerania and his brother-in-law August the younger of Brunswick (Braunschweig-Wolfenbüttel). For the first he formed a famous collection; the second, more parsimonious, nevertheless bought all Hainhofer's papers after his death. To him we owe the survival of a picture — a drawing and some more vivid words — of the episcopal palace and its gardens at the time when the Bishop was preparing a record of his favourite flowers, the greatest *florilegium* the world had yet seen.[1]

Hainhofer's arrangements for his visit were characteristic. He had been visited at Augsburg the previous March by Wilhelm V, who had abdicated in favour of his infant son Maximilian in 1598. The conversation had turned on Philipp II of Pomerania, with whom Wilhelm expressed a desire to correspond. Nothing could be easier, replied Hainhofer, since Philip had asked for copies of drawings in the possession of Wilhelm and the Bishop of Eichstätt. Wilhelm leapt at the opportunity, offered his own and undertook to write to the Bishop for copies of his drawings. The reply that he got and showed to Hainhofer two months later was evidently not intended to be as discouraging as the Bishop's excessive modesty made it sound:

> Highborn Prince, accept my willing and neighbourly service. Gracious My Lord.
>
> As to what Your Highness wrote to me concerning various sketches of diverse animals, plants, herbs or other curiosities and works of art, I may, with the best will in the world, not conceal that I in fact have no such pictures in my possession at present, especially of fish, nor do I have at the moment any four-legged animals, either alive or on paper, except for a few, albeit common birds which for the most part come from Munich. As for various flowers and garden plants, it is possibly no bad thing that some while ago I ordered sketches to be made of what had been observed in my own modest, narrow little garden. However, just now I do not have the drawings to hand but have sent them to Nuremberg, where they are to be engraved in copper and perhaps eventually published in the shape and form Your Highness can see from the enclosed.
>
> However, as to where the *rara* and curiosities mentioned in your letter can be acquired, I can tell you nothing more than that the garden plants were brought back, through the offices of local merchants, above all from the Netherlands, for example, from Antwerp, Brussels, Amsterdam and other places, and were then brought here. Should it please Your

---

1. Philipp von Hainhofer (1578–1647) came from the Augsburg merchant aristocracy on both sides of the family. His father was a cloth-merchant, with premises in Florence as well as Augsburg. The newsletter came to him from his maternal uncle, the correspondent of Henry IV. He had been educated at Padua, Bologna and Siena. A convenient summary of his life, by Friedrich Blendinger, is in the *Neue Deutsche Biographie* (VII, 524–5). The manuscripts acquired by Herzog August and now at the Herzog August Bibliothek in Wolfenbüttel include four texts of Hainhofer's diary of his visit to Eichstätt (MSS. 6.6.Aug.2°, 11.22.Aug.2°, 77.Extra, and 23.3. Aug.2°). The third of these, which also contains poems and letters by Bishop Johann Conrad, was printed by Christian Haütle, 'Die Reisen des Augsburger Philipp Hainhofer nach Eichstätt, München und Regensburg in den Jahren 1611, 1612 und 1613', *Zeitschrift des Historischen Vereins für Schwaben und Neuburg*, VIII (1881), 1–360 (cited here as 'Haütle'). These manuscripts also include drawings (among them the view of Eichstätt reproduced opposite) and engravings, portraits of the Bishop, Philipp II and his wife (their clothes embroidered with naturalistic flowers) and others, and botanical specimens, e.g. hairy grapes and mutant corn, some by the same engravers who also worked on the *Hortus Eystettensis*.

Highness to send someone to our court, I have not the slightest objection but would be glad to have him given an extensive tour of whatever we have here, for I am greatly desirous and concerned to show, to my utmost ability, His Highness and his estimable line every co-operation and service that is pleasant and dear to him, not only in this but in much more. Dated Eichstätt, 1st May 1611.

Your Highness's
Most humble servant,
Johann Conrad.[2]

Diplomatic niceties required a quasi-ambassadorial mission: Hainhofer was armed with a letter of credence, and he set off on 16 May. He arrived next day, sent in his letter and was received with all due ceremony by the Bishop's steward, Wolfgang Agricola, and his Chamberlain, Adam von Werdenstein. The Bishop had hoped to receive him in the garden that afternoon, but was not well enough. Instead, Agricola and von Werdenstein gave him a tour.

> We visited what must have been eight gardens situated round the palace, which lies on top of a hill and bears a fair resemblance to this engraving. Each of the eight gardens contained flowers from a different country; they varied in the beds and flowers, especially in the beautiful roses, lilies, tulips.
>
> The gardens were partly adorned with painted halls and pleasure rooms, in one of which halls stood a round ebony table whose surface and feet were inlaid with flowers and insects engraved in silver.
>
> From the lower castle garden we went through the masons' yard and smithy to the quarry, where we saw the cliff on which the palace stands being blasted with gunpowder and large blocks extracted which were prepared for building, as approximately 200 men from Graubünden and Italians were constantly at work on it. Twelve heavy horses drag the stone uphill.
>
> His Grace wants to turn the whole palace round and have it built from blocks of rock on top of the cliff. It is intended to roof over one side this summer, all of which will be covered in copper; and all together it will cost more than 100,000 florins. The gardens will then be turned round as well and levelled with each other all round the palace on the slope. On the side facing east an exquisite chapel is to be built, with all the windows nine feet high, without any panelling, not even grooves and beams, instead only moulding upon which to hang tapestries.
>
> And by the quarry a stream flows out of the rock which has been diverted round the whole of the palace mountain; it is called the Altmühl and yields excellent trout, pike, bullheads and even nice big crabs.
>
> In the rock of the cliff can be found (fossilised) fish, leaves, birds, flowers and many strange things which Nature makes visible there.
>
> Subsequently we went into the pheasant gardens, of which there are four different kinds: in the one are white pheasants, in the second speckled, in the third and fourth red ones, likewise cranes and other birds.
>
> Just as the gardens are different, so do they also have different gardeners, as none infringes on the other's domain.[3]

On the following day, the Bishop was in better health and able to receive Hainhofer, although indoors. He was duly presented, kissed the Bishop's hand, and made a speech representing Duke Wilhelm's needs. To this the Bishop replied:

> 'I rejoice at the gentleman's presence and dutifully thank His Most Serene Highness Duke Wilhelm of Bavaria, my gracious Lord, for his proffered greeting and the kind confidence he places in me. I wish I had that which His Royal Grace would seek from me. However, since I have nothing His Grace does not already possess, indeed, in more beautiful and better form, and since in addition the flowers, my most precious drawings, are now in Nuremberg (whither I was asked to send them by an apothecary who helped me lay out my garden and increase the number of flowers; he wishes to have them engraved in copper, printed, dedicated to me and to seek his fame and profit with the book), my Lord must suppose that he has caused this trip to be undertaken in vain. However I shall have everything faithfully shown, what little remains of it. Beforehand let us converse a while with each other.'
>
> Then His Grace sat down, pulled a cover over himself and said he could stand no more as his feet were quite unwilling to support him any longer. Thus I had to sit down by His Royal Grace, cover myself as well and remain alone with him for half an hour.

2. Haütle, 19. See full text in Appendix A, pp.60–64.
3. Haütle, 24–6. The engraving is not to be found in any of the surviving versions of Hainhofer's account, but may have been copied from the drawing (reproduced opposite p.1).

Portrait and arms of Basilius Besler and Sebastian Schedel.

> The conversation dwelt on His Highness in Bavaria, his condition and life, on pictures and especially flowers, since His Grace told me that Beseler, the apothecary in Nuremberg, was at that time fully engaged in working on the book; that His Grace would publish it and had one or two boxes full of fresh flowers sent there every week to be sketched; how he always had tulips in five hundred colours, almost all different; and that this book would cost around 3,000 florins.[4]

Next the Bishop showed Hainhofer some of his treasures, making him open a jewelled escritoire 'since he did not have enough strength in his hands to open it himself'.

> Then His Grace rang a small bell. Von Werdenstein and other servants came in and were ordered by the Bishop to conduct me to the balconies in front of his room, to the dressing room and to the treasure chamber. On the galleries, in front of the lovely, bright windows with large crystal lights (so clear that the light shines into the room as if there were no panes and one is tempted to poke one's head out through them), stood various plants: red, yellow, brown and speckled pansies in flowerpots; and in the middle of the gallery violets, apricots, pomegranates, lemons, 'parrot-feathers' (*amaranthus tricolor*), etc. in tubs.
> On this balcony stand six large blocks of the kind butchers use for cutting. In them are placed dead trees and as His Grace was looking out of the window talking to me, I asked him what their purpose was. He answered me that in winter he had a flock of birds in front of his room, since he always had bird food scattered outside and the birds came in masses, sometimes two hundred at one time. They looked for food and sang together. He let them fly free, because if he caught them he would only drive them away and be robbed of his enjoyment.
> A pipe, connected to a tank, was laid through the room and out of the window to irrigate the land beneath. All day long the water runs through the Bishop's own room as if it were a conduit and is conducted up hill through pipes.[5]

So the day passed until dinner, with Hainhofer admiring the many treasures, and putting in a word for his master's needs. At dinner there was a guest, Peter Stewart, professor at the University of Ingolstadt and councillor at Eichstätt. The conversation ran predictably on courts and kings and the desirability of moderation (the Bishop drank mainly water), and also 'about the weather, plants, herbs and many entertaining matters'. Hainhofer offered to initiate a correspondence between the Bishop and the Duke of Pomerania, in whom the Bishop expressed an interest. That night he was not well enough to eat in company, but received Hainhofer again next morning and showed him his coin-collection. So the day passed, and next morning, at 6 a.m., Hainhofer had his final audience. Yet more treasures were displayed, and Hainhofer asked to find an unusually large set of antlers to decorate the new great hall.

---

4. Haütle, 27–8. The Bishop's 'apothecary' was Basilius Besler, and this statement by him is critical to our understanding of their relationship.

5. Haütle, 29. The plants described are all depicted in the *Hortus Eystettensis* (Plates 48, 142, 143, 140, 337).

His Grace leant against the chair with bared head, unable to take a single step, until I had made my last bow at the end of the audience chamber. Then once again he charged me with greetings to His Highness and had me escorted to my chamber to dine on diverse fish. At twelve o'clock the drawbridge was lowered and I was allowed to depart, having distributed tokens of my thanks to the silver waiter, guard, kitchen, cellar and stable staff and whoever else had served me.[6]

All had fallen out as Hainhofer had hoped and planned. Links were established between the Bishop, whose wealth and noble birth were all and more than he had hoped (he was particularly impressed by the proliferation of the Gemmingen coat of arms, even down to the napery with which they were served), and his masters, particularly his favourite Philipp II of Pomerania. They were not to last long, for the Bishop died eighteen months later, but this brief episode clearly meant much to Hainhofer. He retained his friendship with Adam von Werdenstein, and he had his account of his embassy, copied out more than once, together with von Gemmingen's writings, accounts of his family and of the Bishops of Eichstätt, all adorned with engravings and drawings, and now to be found with the rest of Hainhofer's papers at the Herzog August Bibliothek at Wolfenbüttel.[7]

This, and the great book itself, are the only witnesses we have to what must have been one of the most remarkable gardens of its time. It is a pity that the character of the founder remains veiled behind the gauze of diplomatic protocol, since he must have been, as we shall see, a man of an original as well as forceful cast of mind. If pleasure gardens originated in Italy, they were still rare in northern Europe, and there were, as yet, no formal rules for their construction. The Bishop's solution, involving the reorientation of his palace and the construction of eight parterres, separated yet linked by pavilions, was and remained unique. If partly dictated by the terrain and the magnificent situation of the Willibaldsburg above the river Altmühl and the town of Eichstätt below, it was even more required by the desire to display an amazing variety of flowering plants and trees to their best advantage.

We do not know how much the Bishop spent on the exotic new species acquired from the merchants of Amsterdam, Antwerp and Brussels; nor do we know (although we can begin to guess) how much he owed to the friendly help of contemporary botanists. But the *Hortus Eystettensis* remains, as no doubt the Bishop planned, a monument to his passion for beautiful flowers of all sorts, commonplace wild flowers as well as the exotics obtained and cultivated with such care and difficulty. This passion was his most unusual and original trait. Others had grown herbs or vegetables for use, domestic or medicinal; the first scientific botanic gardens had been established 60 years earlier; but Johann Conrad von Gemmingen deserves the credit for acquiring and growing flowers simply as beautiful things, for creating a suitable setting for them, and finally, for causing to be printed the first book devoted to recording their beauty.

6. Haütle, 48. 7. See n.1, above.

THE HISTORY OF THE BOOK

The Bishop and his garden

HE diocese of Eichstätt, one of the oldest in Germany, sanctified by the English missionary St Willibald, had waned during the sixteenth century, its position diminished by comparison with its more prosperous neighbours, Würzburg and Bamberg. Three-quarters of its territory had been lost to the Margravate of Ansbach, the city of Nürnberg, the principality of Pfalz-Neuburg and the upper Palatinate, although the last two had been regained as a result of the Counter-Reformation. Only the Bavarian lands to the south, with the city of Ingolstadt and the small enclave of the Deutsche Orden in the north-west, had been loyal throughout. Previous bishops, notably Martin von Schaumberg who held the see from 1560 to 1590, had tried to introduce the internal reforms imposed by the Council of Trent on clergy and people, but were hampered by lack of capable priests and the need to maintain a watchful policy towards the Protestant north. Matters were not improved by the increasingly independent cathedral chapter, drawn, like their bishops, from the nobility of Franconia and Swabia.[1]

Gemmingen is west of Heilbronn, and the family to which Johann Conrad belonged was 'Uradel', of ancient nobility. Various members of it occupied important positions in the middle ages, and at the turn of the fifteenth century three brothers von Gemmingen were variously Provost and Dean of Worms, Speyer and Mainz, Uriel (1498–1514) becoming Archbishop of the last. The Reformation divided the family, but the branch from Steinegg, on the Würm, remained Catholic, and provided nine canons, two of whom became bishops. The elder, Johann Otto (1545–98), was successively Canon and Dean of Augsburg, but was elected Bishop of Eichstätt by the chapter in 1590 in succession to Martin von Schaumberg. He was unwilling to go, preferring to remain Dean of Augsburg, and in the event became Bishop of Augsburg less than a year later.

Bishop Johann Otto has been characterized as a man of unswerving faith, deep-rooted piety, untiring energy, serious vocation and determined will,[2] all epithets which might be equally applied to his younger nephew, Johann Conrad. He was born in all probability in 1561, but the exact date and place of his birth are unknown; his parents were Dietrich von Gemmingen (1517–86) and Lia von Schellenberg (*d.*1564). He was early destined for the church, being allotted the reversion of a stall by the chapter at Konstanz in 1573. He held a canonry at Ellwangen in 1578–81, but was at Eichstätt in 1579, in Augsburg the following year, and finally took up his prebendal stall at Konstanz in 1588. He matriculated at the university of Freiburg-im-Breisgau on 2 May 1579; on 6 July 1582 the chapter at Eichstätt admitted him to 'first residence', one of the qualifications for the enjoyment of benefices, and in 1583 he took study-leave from Eichstätt at Dillingen University.

He then embarked on six and a half years abroad, nominally spent in study 'utriusque juris'. Between 1584 and 1587 he was in France, mostly at Pont-à-Mousson, including a quarter spent at Paris and, apparently, a visit to England.[3] In 1588 he was in Siena, then at Perugia where on 4 December he became 'Prior of the German Nation' of his fellow students. He went to Bologna, returned to Perugia, and was back in Eichstätt soon after 16 March 1591, when his last permission to study was issued.

Philipp Hainhofer was clearly as impressed by the Bishop's Italianate culture, as by the curiosities he had brought

---

1. This account of the Bishop's life and work owes much to Brun Appel's admirable chapter, 'Johann Conrad von Gemmingen: ein Bischof und sein Garten', in *Hortus Eystettensis: zur Geschichte eines Gartens und einer Buches* (Schriften der Universitätsbibliothek Erlangen-Nürnberg 20, Schirmer/Mosel, München, 1989), 31–68. This invaluable work, cited here as '*H.E.*', also includes Hans-Otto Keunecke's history and catalogue of the exhibition devoted to the *Hortus Eystettensis* at the University Library, Erlangen, in 1989.
2. F. Zoepfl, *Das Bistum Augsburg und seine Bischöfe in Reformationsjahrhundert* (München/Augsburg, 1969), 766.
3. Haütle, 34. The legend that as a youth he was page to Queen Elizabeth is recorded in Andreas Strauss, *Viri scriptis eruditione ac pietate insignes, quos Eichstadium vel genuit vel aluit* (Eichstätt, 1799), 86f, citing a manuscript then at Kloster Rebdorf.

back from his travels. One can sense a reluctance to return, as well as appalled horror at the wars of religion, in the long Latin poem that he wrote to his friend Cleophas Distelmair in December 1587 (the text was preserved by Hainhofer). Only duty, in the person of his uncle Johann Otto, the glory of the house of Gemmingen, brought him home. In the brief months during which Bishop Johann Otto occupied the see of Eichstätt, the deanery became vacant, and it can be no coincidence that the canons elected to it the youngest among them, Johann Conrad von Gemmingen. Soon after his uncle was called to Augsburg, and in his place as Bishop the canons elected Caspar von Seckendorff, already a sick man. A coadjutor was needed, with the right of succession. The canons insisted on their right, not the Bishop's, to appoint, and again chose Johann Conrad. His election was confirmed by Pope Clement VIII with the titular see of Hierapolis, and his investiture was performed by the Dean (and his future successor) Johann Christoph von Westerstetten. He received the temporalities as Prince Bishop from Rudolf II and conducted a visitation with 107 attendants and 81 horses in the autumn of 1594. The following April Bishop Caspar died and on 2 July 1595 Bishop Johann Conrad was consecrated in his place.

There was much to do, within and without his diocese. He needed and got a new Vicar General, Dr Vitus Prüfer, and appointed a new 'praeses', Friedrich Staphylus, who reformed the 'Collegium Willibaldinum', the college for priests founded with insufficient endowment by Bishop Martin von Schaumberg. The increasing confrontation between Catholic and Protestant required constant attention from a frontier state; if his loyalty was never in doubt — he strengthened fortifications and armed his subjects — he did not neglect personal friendship with Protestant princes, to whom on occasion he lent substantial sums. With one in particular, Joachim Ernst, Markgraf of Ansbach, like himself the builder of a famous garden, he kept in close and lively contact. The residence, in danger of collapse, needed urgent attention, and there were other claims on his purse. Despite his evident popularity with the chapter, his relations with them were not wholly smooth, and he took a strong line over the implementation of their 'Wahlkapitulation' (electoral rights) as over excessive hunting. He paid off the debts on the treasury out of his own purse, but was firm in collecting the revenues of his see. He overrode the prohibition of Jews, and took several Jewish families under his protection, but was strong on witch-burning.

He came to Eichstätt rich, and became richer by family bequest. He spent largely on buildings — the cathedral (whose upkeep he shared with the chapter), the other churches of Eichstätt, the Collegium Willibaldinum. He commissioned works of art from all the major craftsmen of his time, notably church plate and monstrances from the goldsmiths of Augsburg and Nürnberg, and vestments from Italy and local weavers. He must also have had a library, but no trace of it has survived.[4] Like all noblemen, hunting (within permitted limits) was a major concern, but his delight in nature was, as we have seen, more broadly based. His interest in coins is evinced in his own handsome coinage, minted for him in Nürnberg.

But his most persistent interest lay in the improvement of his own residence and its surroundings. The immediate needs of the building required the expenditure of 2816 fl., one eighth of the entire household expenditure in 1599/1600. Besides all manner of architectural features (including 'Wappensteine', the cutting of the episcopal arms on stone), were statues and stone benches that can only have been intended for the garden. In April 1599 the Bürgermeister Philipp Jakob Rembold of Augsburg supplied 'certain trees' and 300 asparagus crowns, while the following March Caspar Schmied and Hans Heimhoffer each supplied 100 tubs with stakes — as might be inferred from Hainhofer's account, quite a part of both the garden plants and the trees was movable, depending on the weather.[5]

The most substantial building work undertaken by the Bishop involved a complete change of orientation. Martin von Schaumberg's building had followed the lines of the original medieval fort, facing east towards the fortified gate protecting the entry from the road that led to the castle up the spur on which it stands. As can be seen from Hainhofer's drawing, Bishop von Gemmingen reversed this, so that the Residence now faced west, overlooking the river Altmühl and the town of Eichstätt below. Work probably began as early as 1605, but the main building was

---

4. *H.E.*, 49.   5. *H.E.*, 52–3.

undertaken by Elias Holl in 1609; it was on this that Hainhofer found two hundred men at work two years later. Both the chief mason and the 'Werckh-Maister' were Italian, like many of the workmen.

With this went an equally large relocation of the garden, which had also received Bishop von Schaumberg's attention earlier. The diocesan archives contain the instructions given to two head gardeners, Hans Jeggle, who was given charge at Candlemas 1604 of the '*vor* dem Schloss . . . gelegnen Gärten, sambt dem grossen Perg, an Paumen, mit Zierung der Weinreben, wurtzgarten und allen andern Lustgezierden', and Bernhard Tieget, who in October 1606 was to look after 'alle neue, *hinder* dem fürstlichen Schloss uf St. Willibaldsberg gegen dem Thiefenthal werths gelegne Lust-, Obst- und Kuchen Gärten'.⁶ It is difficult to relate these to the present terrain, the more so since all the buildings were destroyed in the nineteenth century, only a few foundations remaining, but the former seems to relate to the terraces, with a building above, to the north-east of the main building, while the latter, the new 'inner garden', was built round the west end of the hill, between the new façade and the steep fall to the Altmühl. Jeggle's instructions relate largely to the growing of a wide range of vegetables and 'Italian fruits', with special treatment for orange trees, figs, 'Italian and Indian trees', which required different treatment in summer and winter (when they were to be indoors, with windows shut and fires maintained) — evidently, the trees were planted in movable tubs. Unfortunately, since the 'Lustgarten' which presumably contained the flowers and exotic plants was his province, Tieget's instructions are less informative, merely relating to the prevention of entry and supervision of authorized workmen.⁷

There is nothing, in all this, to explain the eight separate gardens that Hainhofer saw, or to illuminate the Bishop's evident passion for flowers. We have his own word for Besler's part in this, and we may guess that Johann Georg von Werdenstein (1542–1608), father of Adam, the Bishop's chancellor, and his friend Joachim Camerarius the younger (1534–1598), a central figure in the remarkable efflorescence of interest in botany around Nürnberg, were further influences, but the only tangible link is the remarkable similarity of the names of the plants engraved on the plates of the *Hortus Eystettensis* to those given in the *Hortus medicus et philosophicus* (1588) of Camerarius. A further indication of their friendship is the signature of Camerarius in the elder von Werdenstein's *Stammbuch* in November 1564.⁸ But, in effect, no further evidence is needed: the *Hortus Eystettensis* speaks for itself.

Johann Conrad von Gemmingen was already ill when Hainhofer saw him, but he maintained his episcopal functions until the last, dying on 7 November 1612; his fine memorial gravestone now stands in the cloister of the cathedral. His successor, Bishop Johann Christoph von Westerstetten, continued his building although with some economies that damaged the original plan by Holl; he also kept up the garden, writing in 1615 to the Pfalzgraf August von Sulzbach of the damage caused by 'die nechstverwichene Langwirige strenge Winterskellt unsern gartengewechsen und Plumenwerckh' — a true gardener's complaint.⁹

Schwertschlager, to whom the history of the Eichstätt garden owes so much, took the romantic view that the garden was sacked and destroyed by the Swedes in 1633-4 during the Thirty Years War, its only survivors four species, now naturalized, that he thought must have escaped from the garden.¹⁰ The truth seems to be somewhat less dramatic. In May 1633, the Swedish troops under Herzog Bernhard von Weimar besieged the Willibaldsburg; the garrison capitulated after ten days and abandoned the Residence. It was captured again in October by Johann von Werth, whose troops entrenched the garden. Damage was no doubt done, but Bishop Johann Christoph's coadjutor and later successor, Bishop Marquard II Schenk von Castell, spent a considerable part of an almost 50

---

6. Jeggle's province was the garden in front of the castle, including the great mount, trees, with charge of the vines, herb-garden and all other amenities; Tieget's all the new pleasure gardens, with fruit and vegetables, lying behind the castle next the deep valley.
7. *H.E.*, 54.
8. *H.E.*, 65, n.115, citing J. G. Suttner, 'Im Mortuarium der Domkirche', *Pastoral-Blatt des Bistums Eichstätt* XIII (1866), 222.
9. He complains of the damage caused by the 'recent long and severe frost to our garden produce and flowers': Joseph Schwertschlager, *Der botanische Garten der Fürstbischöfe von Eichstätt* (Eichstätt, 1890), 41. The letter was quoted from a copy of the correspondence, then in the Amberger Kreisarchiv.
10. Schwertschlager (1853–1924), whose monograph (cited above) was an enlarged version of two lectures given to the Eichstätter Historischer Verein in 1886–7, was professor of natural history at the Episcopal Lyceum at Eichstätt. His book was the first substantial work devoted to the *Hortus Eystettensis*, since the unpublished research of Christoph Jakob Trew (see below, p.21). His main interest lay in establishing the species that owed their naturalization to the garden, and, in particular, those that still survived on the Willibaldsburg and not elsewhere.

years' reign rebuilding and restoring both building and garden.[11] Besler's nephew, Michael Rupert Besler, planning the *Mantissa ad viretum stirpium*, a supplement to the *Hortus*, which exists only in the 1648 calligraphic manuscript at Erlangen University Library, refers in his preface to the restoration of the 'blühendsten und schönsten garten', by Bishop Marquard.[12] In 1640 the Dutch Bollandist, Daniel Papebroch, saw what he described as 'the remains of that famous garden' round the north-west façade:

> It is divided into five connected plots, which surround a beautiful enclosure; the whole is encircled with a crenellated wall, with individual battlements from time to time providing a special place for particular plants or flowers. In the inner part are small wooden pillars, joined to pipes, which bring spring water to irrigate the whole garden. The remainder of what once was part of the garden is now incorporated in the new fortifications.[13]

It is not hard to recognize part at least of the original garden in this changed situation.

With the accession of Bishop Johann Anton I Knebel von Katzenellenbogen (1705–25), not only the garden but the great book took on a new lease of life, if, as we shall see, slow to mature. He was the last prince bishop to beautify and improve the garden, with grass walks in the French style.[14] But more and more it became a garden of shrubs and trees rather than flowers. Bishop Johann Anton I's successors transferred their interest to their town residence, rebuilt, with its own garden, in eighteenth-century style. The old castle remained a garrison, the home of the archives and library, the 'Hof-Bibliothek'. The garden became increasingly the resort of apothecaries, although their needs included many flowering plants. Still, the great botanist and botanical historian Christoph Jakob Trew (1695–1769), to whom the preservation of so much of the history of the *Hortus* is due, visiting the site in 1750, saw 'as well towards evening as at midnight, a quite level and pretty broad flower garden, enclosed with a stone wall'.[15] One senses a wistful *Sehnsucht* for what was no longer there. In 1815 a local lawyer, Franz Xaver Lang, reported exotic plants growing wild on the Willibaldsburg, noting 'Egyptian sage' in particular as a survival from Bishop von Gemmingen's garden.[16] No more has been added to Schwertschlager's list of four other survivors: the snapdragon, the garden violet, the honeysuckle, and the yellow horned poppy.[17]

In 1606, Bishop Johann Conrad's portrait was engraved by Wolfgang Kilian (reproduced above, on the half-title page). He was then 45, and looks somewhat older, though still in good health. The oval frame is surrounded by two pillars with two female figures: that to the left displaying the contents of a treasure chest, with a cherub on top of the pillar disporting among plans and building tools; that to the right holding a vase of flowers in one hand with lily-flowers in the other, and an Agave in a pot at her feet; the matching cherub above holds a cornucopia of fruit and flowers. Below are the Bishop's arms and two devices appropriate to the figures above, representing his main interests in life. It is hard not to feel that the right-hand figure was his favourite.

It may well be that this plate was an early step towards the mass of engraved plates that followed, and was intended (the intention frustrated by the Bishop's illness and death) to appear among the preliminaries of the *Hortus*, as it did in the later edition of 1640. One of the greatest treasures that he caused to be made by Hans Jakob Bair of Augsburg was an elaborate gold and jewelled monstrance. Its design was based on the Tree of Jesse, and incorporated the likeness of the Bishop in the guise of King Solomon, a book in his hand, and his name enamelled on it.[18] Sadly, this great piece was destroyed in 1806, but it must indicate a sense of kinship with the biblical king

---

11. Bishop Marquard describes the ravages of war in vivid terms in his preface to the second edition of the *Hortus Eystettensis* (1646).
12. The manuscript is now part of the collection at Erlangen University Library (MS. 2727, f.10).
13. *H.E.*, 57 and 66, n.141. H. Dussler, U. Kindermann and E. Schubert, 'Daniel Papebrochs Reisebericht von 1660', *Zeitschrift für bayerischer Kirchengeschichte* XLIV (1975), 59–97.
14. A witness in 1716 wrote of 'new and exquisite rarities, statues and a new and handsome garden house', a description repeated about 1730 (*H.E.*, 56–7).
15. University Library, Erlangen, Briefsammlung Trew, Beil. zu Nr. 742-60 (*H.E.*, cat. no.48, and pp.59, 136–7).
16. F. X. Lang, *Topographische Beschreibung und Geschichte der Königlischer bayerischer Kreishauptstadt Eichstätt* (Eichstätt, 1815), 54.
17. Schwertschlager, 18f.
18. Haütle, 42f and 47; *H.E.*, 2, 48 and n 2 and 74 (60, 63).

who shared his love of plants and wrote of them in the great charter of the natural creation between God and himself:[19]

> For he hath given me certain knowledge of the things that are, namely, to know how the world was made and the operation of the elements; the beginning, ending and midst of times: the alterations of the turning of the sun, and the changes of the seasons: the circuits of the years, and the positions of the stars: the natures of living creatures, and the furies of wild beasts: the violence of winds, and the reasonings of men: the diversities of plants, and the virtues of roots.
> And all such things as are either secret or manifest, them I know. (*Wisdom of Solomon* VII, 17–21).

Like Solomon, too, the Bishop was famous and sought out for this particular branch of wisdom.

> And Solomon's wisdom excelled the wisdom of all the children of the east country, and all the wisdom of Egypt . . . And he spake of trees, from the cedar tree that is in Lebanon even unto the hyssop that springeth out of the wall: he spake also of beasts, and of fowl, and of creeping things, and of fishes.
> And there came of all people to hear the wisdom of Solomon, from all kings of earth, which had heard of his wisdom. (I *Kings* IV, 30, 33–4).

The analogy between the god-like heroes of the past, even with God himself, was not lost on such heroic figures of the counter-reformist church as Pope Sixtus V.[20] It was one which Bishop Johann Conrad accepted, in his portrait, in the monstrance, and, as we shall now see, in the great book that recorded his purpose in making his garden.

Printing and publication: the evidence of the book

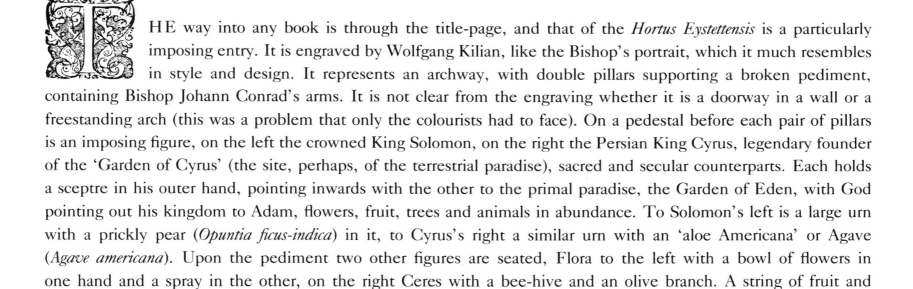

THE way into any book is through the title-page, and that of the *Hortus Eystettensis* is a particularly imposing entry. It is engraved by Wolfgang Kilian, like the Bishop's portrait, which it much resembles in style and design. It represents an archway, with double pillars supporting a broken pediment, containing Bishop Johann Conrad's arms. It is not clear from the engraving whether it is a doorway in a wall or a freestanding arch (this was a problem that only the colourists had to face). On a pedestal before each pair of pillars is an imposing figure, on the left the crowned King Solomon, on the right the Persian King Cyrus, legendary founder of the 'Garden of Cyrus' (the site, perhaps, of the terrestrial paradise), sacred and secular counterparts. Each holds a sceptre in his outer hand, pointing inwards with the other to the primal paradise, the Garden of Eden, with God pointing out his kingdom to Adam, flowers, fruit, trees and animals in abundance. To Solomon's left is a large urn with a prickly pear (*Opuntia ficus-indica*) in it, to Cyrus's right a similar urn with an 'aloe Americana' or Agave (*Agave americana*). Upon the pediment two other figures are seated, Flora to the left with a bowl of flowers in one hand and a spray in the other, on the right Ceres with a bee-hive and an olive branch. A string of fruit and vegetables hangs down from each corner of the pediment.

The upper part of the arch is filled with a cloth (in the earlier copies always coloured yellow, so evidently a cloth of gold), and upon it the title, explained as a 'diligent and accurate delineation and representation to the life

---

19. The parallel with Solomon is further emphasized in the title-page and Besler's dedicatory letter in the *Hortus Eystettensis*. See also Werner Dressendörfer, 'Vom Kräuterbuch zur Gartenlust', *H.E.*, 84 and 89 n.45.

20. G. Cipriani, *Gli obelischi egizi: politica e cultura nella Roma barocca* (Florence, Olschki, 1993), 35ff.

of all the plants, flowers and shoots, collected from various parts of the world with singular care, which are now to be seen in the gardens surrounding the episcopal citadel'. All this, it says, 'Opera', achieved — by whom? The next two lines give us the answer: it is the work of Basilius Besler, lover of medicine and apothecary; finally comes the date, 1613. The last three lines below the word 'Opera' are off-centre and look like an after-thought.

The modern reader, used to more functional title-pages, may ask what role is claimed for Besler. Is he the author? Or the publisher? The question is not easily answered, nor is it made clear by the preliminary pages, the dedications (for there are two), the addresses to the reader, that follow. Cataloguers usually claim him as the author (but to a cataloguer an author is more necessary than a publisher). The true answer seems to be 'betwixt and between', or rather that what the Bishop told Hainhofer was entirely accurate, namely that Besler, who had helped him form the garden, had wished to have the Bishop's drawings engraved and printed with a dedication to the Bishop, and 'to seek his fame and profit with the book'. This would have been a perfectly normal relationship between a promoter and a noble and rich patron, and all might have gone well if the Bishop had not become ill and died before the work was completed.

Besler was clearly plunged into difficulties by this misfortune. There were immediate difficulties with the publication of the book, and others, longer lasting. The new Bishop was prepared to finish his predecessor's work, in this as with his building, but as a duty rather than, as heretofore, a shared pleasure. There were inevitable misunderstandings about money. There were also, it seems, other misunderstandings with colleagues who had helped Besler, for which he was blamed by their successors. Finally, modern bibliographers, trying to reconcile documentary references with the physical evidence, the surviving copies, that is, of what appear from their title-pages and preliminaries to be five separate editions, have again depicted Besler as a rogue, depriving the dead Bishop of his due by pirating his work.[1]

In all this Besler's voice has not been heard, nor can it be now. It is certainly clear that there were, at the outset, two separate printings, but it is less clear that one of them was piratical; misunderstanding seems a more likely explanation then fraud. Baier, who first attempted to distinguish the two printings, found a simple difference: both had the same title-page, but those with the dedication to the old Bishop, Johann Conrad, were on unwatermarked (but superior) paper and were legitimate; those dedicated to the new Bishop, Johann Christoph, were on watermarked paper (apparently inferior) and by their later date clearly convicted as piracies. There are a number of difficulties in this account, which require examination. First, why should the 'later' copies, if piratical, be dedicated to anyone, let alone a live Bishop who might well take offence at the misuse of his name? Secondly, the connection between each dedication and the presence or absence of the watermark is not so absolute as Baier supposed. Sheets without a watermark appear in some copies with the dedication to Bishop Johann Christoph; more significantly, one of the drawings for the plates, now at Erlangen University Library (f.107 = plate 89), is on watermarked paper, which proves that it must have been available at this stage, not exclusively after Bishop Johann Conrad's death. Furthermore, the unwatermarked paper is far from uniform: several different stocks are found, some thicker, some thinner. Similarly, several different forms of the watermark, a large cartouche enclosing a bunch of grapes on a cup,[2] are visible in all the copies in which watermarked paper is used. Again, there is an important distinction not noted by Baier: some copies (all on unwatermarked paper, and all with the dedication to Bishop Johann Conrad) have the text printed on the reverse side of the plates; others, some watermarked, some not, and diversely dedicated, have no printing on the versos, and the text (if present) printed on both sides of a sheet interleaved between the plates. Finally, of this latter group, the majority have been coloured by hand, while only a few of the copies with text backing the plates have been so coloured.

---

1. The most thorough account (but see above, p.5, n.1) is that of Hans Baier, 'Die Ausgaben des Hortus Eystettensis 1613–1750', *Aus dem Antiquariat* XCV (27 November 1970), 273–280. Other descriptions are in Claus Nissen, *Die botanische Buchillustration*, 1951–2, I, 70–3 and II, no.158; E. Schmidt-Herrling, 'Hortus Eystettensis: vom bischöflichen Garten zu Eichstätt', *Frankenspiegel*, II. 5 (1951), 8–9; and Theodor Neuhofer, 'Basilius Besler – Hortus Eystettensis', *Historische Blätter für Stadt und Landkreis Eichstätt* V (1958), 20–1. The 1713 edition was described by Allan Stevenson, *Catalogue of botanical books in the collection of Rachel McMasters Hunt* II (1961), 46–9.
2. C. M. Briquet, *Les filigranes*, ed. A. Stevenson, 1968, I, 2122, where it is listed under 'Armoiries: Pomme de Pin'. Several different forms of the watermark are to be found.

There is, as we shall discover, a possible explanation for all these peculiarities. Meanwhile, the sheer size of the enterprise must strike us. The actual dimensions of the book were determined by the size of the paper used, called 'royal' (and traditionally watermarked with a bunch of grapes), measuring about 22½″ × 18″ (570 × 460 mm).[3] It was the largest then available, and it was used broadside, unfolded, the leaves oversewn together to form gatherings for binding. The bulk of the book was made up by the 367 plates, for which no less than ten different engravers were employed.[4] Together with the preliminary pages, each copy consists of between 388 and 576 leaves, depending on whether the text backed the plates or was printed separately. The text required over 400 separate letterpress workings; the plates 370 runs through the rolling press. The number of copies printed was, according to contemporary documents, at least 300; of these many, both coloured and uncoloured, have survived.

The co-ordination of the two separate printing processes to produce two separate editions, each with distinct demands on each process, was no light task. The costs of the labour and materials — engravers, printers, copper plates (copper was not cheap) and the different stocks of paper — were enormous. The distinction between the two editions lies not in the dedications, but in the paper stocks and the printing of the text. The common edition, uniformly dedicated to Bishop Johann Conrad, has the text backing the plates and is printed on unwatermarked white paper (±32 mm between chainlines), relatively light, given its size, and of variable thickness. The less common copies, which have the versos of the plates blank, and the letterpress text either absent or printed on both sides of leaves inserted between the plates, are printed either on even, slightly creamy paper with the grapes watermark (±30 mm between chainlines) or on unwatermarked paper, like the common edition, some sheets noticeably thicker than others.

The purpose of this distinction is clear. The copies without text on the back of the plates were intended to be coloured, and the versos left blank deliberately in case the colours permeated the paper. The majority of coloured copies are printed in this way; those with text on the versos are, for the most part, demonstrably late. Only four copies with blank versos to the plates are uncoloured; three, significantly, have no letterpress text, and the fourth has it on separate sheets.[5] At least one more such copy must have existed until recently, since some (if a small proportion) of the single sheets from the *Hortus Eystettensis* offered for sale today, in modern colouring, are from a copy (or copies) with the plates unbacked with letterpress. Further, among the preliminaries, the dedication to Bishop Johann Conrad and the addresses to the reader are on unwatermarked paper, while the dedication to Bishop Johann Christoph and the privileges are on watermarked paper; the engraved title and the page with Besler's portrait and arms vary. There may be some significance in these divisions, but it must be remembered that, with each leaf a separate sheet, there is no integral connection between any of these and the main part of the book, in either form.

The double printing of text and plates of so large and long a book took, as we shall see, many months. The different paper stocks can only have been worked off simultaneously, presumably beginning with the superior unbacked copies, but with the other following immediately, sheet by sheet; the alternative, printing each paper stock separately, would have almost doubled the length of an already lengthy process. The existence of the two editions must, then, have been pre-arranged, part of the plan arranged between Bishop Johann Conrad and Besler. We have no means of knowing whether it was intended to reflect some division of interest, of investment and eventual return, but it seems unlikely. The Bishop certainly bestowed large sums of money on the venture, but it would be quite out of character for a wealthy and noble patron to expect a financial return; indeed, he was quite explicit in what he said to Hainhofer — Besler was to seek fame and fortune, while his return came in the coinage appropriate to a patron, a dedication couched in suitably honorific terms.

What, then, was the function of the two editions? That on plain paper, with text backing the plates, was, we may surmise, for the trade, the established book trade, with which it was Besler's job to deal. There is some evidence

---

3. Papers of this size are recorded from the 14th century onwards (see P. Gaskell, *A new introduction to bibliography*, 1972, 67–73). The common French name for it was 'grand raisin'.
4. The signatures of ten engravers appear on different plates (see below, pp.29–30, and Appendix D); many of the plates are unsigned, however, and may be by other hands.
5. These copies are at Bonn (Universitätsbibliothek), Frankfurt-am-Main (Stadt-und Universitätsbibliothek), Marburg (Universitätsbibliothek) and Augsburg (Staats- und Stadtbibliothek).

that he expected, and had, some difficulties: the marketing of so vast a book, even at what was in effect a subsidized price, would have stretched the resources of a trade increasingly used to smaller books. But even here the Bishop was, it seems, prepared to help. Besler's 'fame and profit' were to come from the 'plain' copies. In the event, these sold well, and remarkably quickly, to those who clearly prized them, then and since. There are a large number of copies of the plain edition of the *Hortus Eystettensis* still left in the world, in every major public and private library, even though the number has been sadly diminished by the modern habit of breaking such large books for the greater profit from selling each plate separately. Ultimately, then, this part of Besler's sales campaign must be accounted a success.

The fate of the copies destined for colouring is more complex, and it may well be that the Bishop had a more direct interest in these (although still not a financial one), so that his death did indeed require a change of plan. These copies were designed for the wealthy and great, people like Hainhofer's patrons. Like Wilhelm V of Bavaria they were less interested in scientific details than in 'contrafetten', images that they could possess of the wonders of nature. This may explain why these copies are usually without text; some have the common German names added in manuscript, others have the full printed text but printed on separate sheets. The colouring of such copies was an enormously expensive business, but the Bishop clearly intended to give some as presents, and his support would have encouraged the sale of others. His death, therefore, brought about some of the difficulties that Besler now had to face. In some cases, the coloured copies are dated by the illuminators; the dates show that they were hard at work over the next two years and more. It may well be that financing this work and collecting its price (coloured copies seem to have cost ten times as much as plain ones) was an extra and unexpected burden. Hainhofer provided some help, if not very effectually. The continued provision of such copies was clearly quite distinct from the marketing of the plain copies, and it was to last for many years. What distinction was intended by the two different papers, unwatermarked and watermarked, is hard to fathom now. Those dedicated to Bishop Johann Conrad and on plain paper are, in some but not all instances, demonstrably earlier. But the watermarked paper (superior, not inferior as Baier supposed) was at least available as early, even if the dedication to Bishop Johann Christoph suggests a later issue. There are considerably fewer of these, and the existence of the two outstandingly well coloured copies on this paper suggests that it was intended for *de luxe* copies.

There were further complications besides the expense and organization of the work. First of all, there was the printed text. If the names originally given to the plants (and engraved on the plates) owed much to the younger Camerarius, he had been dead for fourteen years.[6] Although Besler had come to take his place as the Bishop's confidant in garden matters, he was not a professional botanist but an apothecary and a collector of natural curiosities. The other works that he printed relate to these interests. The nomenclature of plants was a serious and difficult matter.[7] The printed text was intended to be an elaborate contribution to it, with its careful descriptions, common names and, above all, cross-references to previous herbals, Dodoens and Mattioli, Bauhin, Clusius (Charles de l'Ecluse) and Lobelius (Matthias de l'Obel), Camerarius himself, and such recent works as Schwenkfelt's *Stirpium et fossilium Silesiae catalogus* (1600) and even Castore Durante's *Hortulus sanitatis* (1609). All these are listed, and more, under the heading 'Autores, qui in opere Horti Eystettensis citantur' in the preliminary pages. There is a persistent tradition, first printed in the eighteenth century but quoting contemporary sources, that much if not all of this was the work of Ludwig Jungermann (1572–1653), the nephew of Joachim Camerarius the younger, work unacknowledged by Besler.[8] If so (and the number of errors in the text suggests that Jungermann can have been only a partial contributor), it did not prevent an enthusiastic encomium from the Dean of the Collegium Medicum of Nürnberg, of which Besler was a member, which also appears in the preliminaries.

---

6. Camerarius died in 1598. Next year Bishop Johann Conrad began enlarging his garden. Besler dates his own involvement in the work from 1607.
7. Hieronymus Bock echoed a widespread feeling when he wrote of the 'Bibernell' in his *Kreuter Buch* (Strassburg, 1551), f.177a: 'Hilff Gott, was hat dise gemeine wurtzel sich müssen leiden bei den Gelehrten; haben alle darüber gepümpelt und gepampelt, doch nie eigentlich dargethons wie sie beiden Alten heiss oder was es sey'.
8. See below, p.18–19.

The first dedication is a fine triumphal paean to the Bishop, echoing the message of the title-page, with the same references to Solomon and Cyrus; it emphasizes the labour, the long journeys, that have brought the garden to fruition, and its semblance, the book, in which all, spared this travail, can now share. The second dedication to Bishop Johann Christoph von Westerstetten is rather more than an expression of gratitude for his continued patronage of what his predecessor had begun. Besler speaks of the greatness of the work achieved by his predecessor's generosity, and his public duty to complete it. Yet he fears it may find few buyers. Given all his work on the plates, engraved to the life, the descriptions he has compiled, the huge cost of it all ('I forbear to mention the engravers' wages'), it would be a crime to hide such a light under a bushel, as all good judges agree; the public demands that it be published for all to see, nor should they be burdened by too high a price. He emphasizes the novelty of the work: where others have given 'nudas herbarum imagines', bare copies, diminished, he has given them as they really are, and life-size. There is a subdued note of defence in all this, which now changes to a ringing cry of confidence in the new bishop, worthy successor of his old patron, heir also to a garden without equal, to whom by every right this dedication is due. The difficulties that were to ensue with the Bishop, or rather with the Chapter of Eichstätt, lay in the future. There was much — the existence of two editions, the ownership of the copper plates — left unstated or obscure in Besler's dealings with Bishop Johann Conrad. The conversion of a shared enthusiasm into a business transaction strained a hitherto firm if tenuous understanding. We do not know what arrangements for publication had been originally intended, only that Besler had to shoulder the business without his patron's support.

But the *Hortus Eystettensis* was a triumph, not only for Bishop Johann Conrad, but also for Besler. If the title-page represented the Bishop's view of the garden and the book as a living or pictorial version of Psalm 150, a tribute to the Creator of all things, into which the intrusion of Besler's name was apparently an afterthought,[9] Besler's own view is made clear in two other engraved plates that precede the text. The first is simple enough: the representation of Besler himself, soberly but richly dressed, and holding a sprig of basil, *Ocimum basilicum*, in his hand, with, beside and on a separate copper, the arms of his family; the portrait is dated 1612, and describes Besler as 'rei herbariae studiosus'.[10] Beneath are some commendatory verses by Besler's colleague Georg Rem. (Reproduced above, on title-page.)

The other plate, signed by the Nürnberg engraver Leypolt, is stranger. It was to serve as the part-title for each of the four seasons which formed the main divisions of the *Hortus Eystettensis*, and the middle is empty, so that a separate title plate, one for each season, can be printed within. The rest of the plate is like, yet unlike, the title-page. There are two pillars to left and right, but if one has a plant with various seed pods coiled round it, the other is encircled by a snake, with various lizards adjoining. Above are two female figures looking down, one the familiar figure of Flora, but the other 'Salus' with a bottle labelled 'Elixir' in one hand, and the other resting on a pestle and mortar. The foreground shows not the garden of Eden but a pyramid of medicine bottles, sundry retorts and vessels and four trays of metals, minerals, gems and other stones. Still more strangely, the oval frame and Besler's arms, repeated triumphantly in the centre above, are quite clearly enclosed within tortoise-shells, that of the frame further embellished to left and right at the top, and at the foot, with grotesque heads.

All this accords rather oddly with the rest of the *Hortus Eystettensis*, but it could hardly be a more telling commentary on the personality and interests of Besler himself. The self-confident, armigerous, citizen of Nürnberg, proud of his calling as a chemist, a member of the guild of Apothecaries — this was the man who promoted and carried through the enormous task of giving the *Hortus Eystettensis* in published form to the world.

---

9. It is impossible to guess how the title-page would have ended if Bishop Johann Conrad had lived to see his vision realised. It is clear, however, from the asymmetry of the last three lines that they were added later, when Besler assumed a role and responsibility for the work not originally intended.

10. What may be the original (if it is not a copy) of this portrait is the painting in oils on copper, measuring 275×205 mm, in the Gothenburg Art Gallery. Like the print, it is surrounded with an oval inscription, the wording identical. It is signed 'LS 1612'; the painter Lorenz Strauch (1554–1630) lived and worked in Nürnberg (Thieme-Becker, *Allgemeines Lexicon der bildenden Künstler*, XXXII, 170–1). I am grateful to Gunnar Göthberg, apothecary at Göteborg, for informing me of this picture.

Printing and publication: the archival sources

ALTHOUGH books themselves do not tell us all that we would wish to know about their manufacture and subsequent history, archival references can be even more provoking. The light they provide can be bright indeed, but it is often fitful and sporadic: the darkness in between can be the darker. In particular, the difficulty of fitting the facts presented in the archival sources to those derived from study of the *Hortus Eystettensis* itself suggested that the two should be treated separately. What follows, then, is an attempt to fit the references, given in full below (Appendix A), into the framework of the printing and publication of the book already constructed. They are taken, as presented, in chronological order, with one important exception, namely the summary of facts relating to the book and Besler's obligations compiled for the Chapter of Eichstätt, probably in 1630, soon after his death. (The original of this document is no longer to be found, but it exists in a transcript made in 1751 by Johann Georg Starckmann, of whom more later).[1] As this relates to other documents, and is itself a chronological compilation, its facts have been taken as and when relevant. The other sources are Hainhofer's travel diary and correspondence, the minutes of the Nürnberg city council (the 'Rat') and the Eichstätt chapter, and the preliminaries of the *Hortus Eystettensis* itself.

It is the last that provides the first date, 1607. 'Six years ago', wrote Besler in his letter to the reader, the work, 'Opus botanicum', was begun. It is uncertain, in the context, whether this refers to the building of the garden and his own involvement in it, or to the drawing of the flowers and the scheme for a great printed work. As the builders were still busy on the site of the 'inner garden' the previous year, the earlier of these alternatives is the more likely.

Four years then pass, until 15 April 1611, when the Nürnberg Council first took notice of the enterprise.[2] The arrangement to which it gave assent is somewhat obscure, but it looks as if the Bishop, at the instance of Besler, was providing credit, or at least credibility, for the enterprise, at the same time drawing the Council formally into the enterprise. The Council had also to ratify the employment of craftsmen. If Besler were now negotiating with several engravers, all resident in Nürnberg, he may well have wished to have his negotiations underwritten by the Council, for which the Bishop's offer of cheap corn was no doubt an acceptable exchange. It may be, too, that this was a form of insurance for the Bishop in the event of fraud or misadventure on the part of Besler.

It was very shortly after, on 24 April, that Besler dated his dedication to the Bishop, and a week later that the Bishop wrote to Wilhelm V, enclosing what was evidently an early proof of one of the engravings.[3] Eighteen days later, Hainhofer was in conversation with the Bishop. For the first time, Besler's name is mentioned; the Bishop's pride in his tulips ('five hundred fold' is a tolerable exaggeration) is evident. Perhaps the most interesting fact is that boxes of flowers were sent to be drawn in Nürnberg. Some at least would hardly stand the day's journey thither, and it may be that they were only to provide detail to supplement the rough sketches, 'ad vivum', made on the spot. 3000 florins was to prove a considerable underestimate of the cost.

Less than a month later Hainhofer was writing to his other patron, Philipp II of Pomerania, again enclosing proofs of engraved pages.[4] The desirability of a 'proper, illuminated' copy shows that colouring was certainly envisaged already at this stage of the enterprise. On 25 June he is able to promise that the Bishop will give Philipp a copy: as they had never met, this is an indication of the lavish scale of the Bishop's intended generosity, or of Hainhofer's optimism. The further progress of the book then rather faded from Hainhofer's promotional enterprise

---

1. The document, headed 'Acta und Inhalt des Eystättischen Gartenbuchs, so sub regimine Herrn Bischoff Conrads von Gemmingen zu Nürnberg in druck gegeben und angefangen worden. Anno 1612', is so described in the Staatsarchiv Nürnberg, Bestand Hochstift Eichstätt, Literalien Nr. 98, Abt. 22, Nr. 3. Trew's transcript, headed 'Extractus', is in Erlangen University Library, Briefsammlung Trew, Starckmann Nr. 37. See *H.E.*, 93.
2. Th. Hampe, *Nürnberger Ratsverlässe über Kunst und Künstler*, Leipzig/Vienna, 1904 II (1571–1618), nos. 2388, 2520, 2528, 2587 (see Appendix A, p.64).
3. C. Haütle, 'Die Reisen des Augsburger Philipp Hainhofer in Eichstätt, München und Regensburg', *Zeitschrift des Historischen Vereins für Schwaben und Neuburg*, VIII (1881), 1–360.
4. Oscar Doering, *Des Augsburger Patriciers Philipp Hainhofer Beziehungen zum Herzog Philipp II von Pommern-Stettin*, Vienna, 1894. (See Appendix A, p.63).

for a year; when he mentions it again, it is only by way of explanation for the non-delivery of some paintings on vellum, the artist, Daniel Herzog, being occupied with the *Hortus Eystettensis*.[5]

Between May and July 1612 Besler obtained the three privileges for the *Hortus Eystettensis*; no such protection was available in a divided Germany. The following August Hainhofer reports again on the progress of the book, again indirectly. Georg Gärtner has just finished an engraving of a 'mayen krueg', a vase of flowers, for the Markgraf of Ansbach, a *tour de force* of his skill, which is also being deployed in the *Hortus Eystettensis*; a coloured copy is to be found for his noble correspondent.[6]

In October and November, the Nürnberg Council is again preoccupied with Besler's affairs. Balthasar Caimox has done him some wrong (it may be in connection with the distribution or marketing of the *Hortus Eystettensis* at the Frankfurt book fair), and Besler is trying to persuade the local Council to exert some pressure on its Frankfurt counterpart.[7] It may be some such eventuality that the Bishop (not now long to live) and Besler foresaw when they involved the Nürnberg Council in the enterprise the previous year.

After Bishop Johann Conrad's death on 7 November, there is a long gap until the following July, by which time the *Hortus Eystettensis* has been finished. A copy has been presented by Besler to the Nürnberg Council, which they wish to return, demanding a coloured copy for which (rather surprisingly) they expect to pay. On 24 August 1613 Besler dates his second dedication to Bishop von Westerstetten, and on 23 September he rendered an account of some sort to the Chapter at Eichstätt (see below). On 12 October the Chapter directs Adam von Werdenstein, perhaps now Besler's intermediary, at least someone with intimate knowledge of the previous Bishop's plans, to deliver to it 25 copies 'primae correctae et perfectae editionis'. The Latin phrase surrounded by German suggests a quotation, perhaps from some earlier document, a formal note in the Chapter's 'Acta', for example, and it may be that the confusion caused by it, then and since, stems from this fact.

But what they wished to imply by this direction has been obscured by the passage of time. The Latin phrase so exactly echoes the traditional (but more recent) formula 'first and best edition', the standard encomium in the second-hand book trade, that Baier and others before and since have taken it to mean just that. In this case, the assumption would be that they wanted copies, on the best paper, with the early and superior state of the plates, but no text, rather than the 'plain' but complete edition with the printed text backing the plates. It seems more likely, however, that the word 'editio' is to be taken in its literal sense of 'offering for sale'. All the evidence suggests, as we shall see, that the plain copies were put on sale first, and that the copies with the plates unbacked with text and destined for colouring were not ready until later.

In this case, 'correctae et perfectae' can be interpreted in the sense in which it has hitherto been taken, namely that the copies required were to be 'perfected', with the letterpress text printed on the other side of the sheet to the plate. The 25 copies were clearly needed for each member of the chapter (a demand that was to recur a century later in the history of the book). It is not certain whether they got them or not, but, as there is no evidence of subsequent complaint on this score, the probability is that they did.[8]

For complaint there was, and it is time to consider the Eichstätt Chapter's list of grievances, the 'Extractus' from the Acts of the Chapter prepared *c*.1630, which in turn involves going back to an earlier stage in the production of the book. The first document cited is a memorandum by Besler dated 1612, and, by implication from clause 3 of the list, written after Bishop Johann Conrad's death, probably soon after, since he died on 7 November. This

---

5. Daniel Herzog was regularly employed by Hainhofer as a flower-painter throughout this period (1610–15). See Thieme-Becker, *Allgemeines Lexicon* XVI, 561. He may have also been employed as a colourist for the *Hortus Eystettensis* (see below, p.36).
6. This print by one of the main engravers of the *Hortus Eystettensis* is known in two states, one with a dedication to Joachim Ernst, Markgraf von Brandenburg, signed 'Georgius Gärtner, Norimb. Pictor A. 1612', the other with six lines of elegiac couplets on the theme 'all flesh is as the grass', beginning 'Nunc nitidi redolent flores sed tempore parvo/Durat odor fumique instar abire solet.', signed 'C.H.' This may be Caspar Hofmann (see below, p.18). The print is only found in a few copies of the *Hortus Eystettensis*, and is not part of the regular contents. See Baier, 275 n.6.
7. Balthasar Caimox (*c*. 1583–1635), print publisher and art-dealer at Nürnberg. See Thieme-Becker, *Allgemeines Lexicon*, VI. 241–2.
8. The copy of the *Hortus Eystettensis* now at the Economisch-Historische Bibliotheek, Amsterdam, may well be one of these copies. It is inscribed at the foot of the titlepage 'Reverendissimo in Christo Patri ac Dño Dño Thomae electo Abbati in Salem Dño suo gratioso atque observantissimo hunc codicem obtulit et donavit Clarissimus Dñs Bartholomaeus Richel ex Neufra J. V. Licens. et R^{mi} ac Ill^{mi} Principis ac Dñi Episcopi Eÿstettensis Vice Cancellarius. Ex Eÿstadio die 13. februarÿ A.º MD.CXV.'

recites the arrangement between the Bishop and Besler, in terms now familiar, but reveals some interesting further facts. Of the plates, correctly numbered as 370, '50 were prepared in Augsburg at the order of the Bishop'; these were presumably all executed at the Kilian workshop (Wolfgang Kilian, as we have seen, was the engraver of the Bishop's portrait). At the Bishop's death, the work was 'scarcely half completed'. This may refer to the engraving of the plates alone (in which case it would imply that not more than 185 were finished), or (perhaps more likely) to the task of printing as well, in which case it is highly probable that more than half the plates were engraved. The centre of operations must in any case have switched to Nürnberg, where most of the other engravers worked,[9] and where the letterpress and plate printing workshops were no doubt also situated, conveniently under Besler's eye.[10]

By this time, too, a total of 7500 florins had been advanced by the Bishop 'for this work'. This presumably subsumes the 3000 florins mentioned by the Bishop to Hainhofer eighteen months earlier. There is no way of telling exactly what part of the costs it was intended to cover. It is unlikely that it was spent on the original drawings, which the Bishop most likely paid for himself (although, as we shall see, the process by which an 'ad vivum repraesentatio' was converted into the engraver's image was a complex one). If it covered half the plates, it works out at about 40 fl. a plate, which seems too much. It is probably best regarded simply as an advance of working capital to Besler. It is significant that there is no suggestion that this was a loan, or that Besler was required to do more than account for its expenditure.

We shall return to this document, but in the meantime the post-publication history of the *Hortus Eystettensis* can be followed in Hainhofer's correspondence. By 18 December 1613, the prospect of the gift of a coloured copy for Philipp II has receded, and one can only be obtained at a price of 500 florins. The same price is reported (in Italian) to Herzog August at Braunschweig, together with the offer of a plain copy at 35 fl. He orders a copy to be sent from the Frankfurt book fair, but is frankly incredulous of the price of colouring — can he have read the price correctly or should it be only 50 fl? This *contretemps* seems to have taken some time to disentangle, and it is only in the following June that Herzog August is giving instructions for binding; the difficulty of coping with a book printed in single sheets is manifest, and his solution, to treat it as an atlas with the sheets mounted on raised guards, was a sensible one. There was still to be a confusion over the collation — a frequent problem with books of any size, and especially so with this — but it seems to have been resolved. A year later, on 14 October 1615, Herzog August orders another copy for his brother-in-law, the Graf von Oldenburg-Delmenhorst.

By now the dispute between the Chapter at Eichstätt and Besler seems to have broken out.[11] Clauses 4 to 7 in the 'Extractus' cite a whole range of different documents or references in the Chapter's 'Acta', all apparently dealing with the consequences of maintaining Bishop Johann Conrad's enterprise. The first of these, significantly, is entitled 'Contract de Anno 1615', which suggests that there had been no previous such document, and that, like his predecessor, Bishop Johann Christoph von Westerstetten had allowed matters to run on without a binding contract. There is, however, no mention of a further advance, merely the record of a first impression of 300 copies, plus the grant of permission for Besler to retain the plates for four years and to use them for his own purposes. The next three clauses are obscure and contradictory. They complain, with evident justice if he did so commit himself, that Besler failed to return the plates after the four years had elapsed, but also that he printed another edition, which, however, seems to have been envisaged and allowed by the 1615 contract quoted in clause 4. It is difficult to resist the conclusion that the Chapter was aware that there were two editions; of one they had, in all probability, been provided with copies, but about the other they had grave but unspecific fears that Besler was somehow illicitly profiting from it. No such edition, clause 6 is emphatic, have they been involved in. It is theoretically possible that

---

9. There were, in fact, close links, in some cases family relationship, among the two groups, the Kilians in Augsburg and those who worked in Nürnberg. See Nissen, *Buchillustration*, I, 71–2.
10. Baier (275 and n.8) attributed the letterpress printing to Konrad Bauer of Altdorf, on the strength of a similarity between the types and initials used in the *Hortus Eystettensis* and in L. Jungermann *Catalogus plantarum* 1615. Jungermann, he further noted, was congratulated by a colleague on 28 September 1613 on the *Hortus Eystettensis* since 'per te typis mandatus est'. Keunecke (*H.E.*, 95) rightly points out that these grounds for identification are insufficient, and that it is unlikely that a small printer in Altdorf would have been capable of so large a production as this.
11. At least four separate documents dated 1615 are cited in the 'Extractus', which suggest that a case was being prepared for negotiation that led to the 'Contract'.

Besler over-ran the letterpress sheets (which were certainly not set and printed again in the seventeenth century) and reprinted the plates (the re-use of which was clearly the *casus belli*), but there is no sign of it at this stage.

Then comes a curious 'contra' entry in the Chapter's charge, that the coloured copy sent to Eichstätt (the word 'sole' probably refers to 'coloured', and does not imply that it was the only copy of any sort) cost 500 florins to produce in labour alone, and that this was confirmed not only by the 'account' rendered on 23 September 1613, but also by the price paid for a similar copy by the Nürnberg Council, borne out by its minute of 2 July 1613 . The document is not, then, a one-sided attempt to sum up the charges against Besler; this is evidence that his statements about the costs he had incurred are genuine and deserve consideration.

All this, it must be emphasized, is a summary of documents made fifteen years after the event by someone unfamiliar with the facts and events described. If at pains to reach a conclusion, the writer can only list what he finds in the 'Acta', which (like those of the Nürnberg Council) convey what was considered important at the time without thought for the needs of posterity. How far in the past these events were is vividly illustrated by the next extract from the correspondence of Herzog August and Hainhofer.

On 11 January 1617, Herzog August, obviously satisfied with the effect of his last gift, wrote and asked Hainhofer for another copy of the *Hortus Eystettensis*. The answer must have surprised him. Hainhofer writes on 9 February that he has bought a copy and, wisely, since the book would have been in single sheets, sent it to be sewn. It lacks some leaves, for which he has written to Besler, and — this must have startled the Duke — the book itself was rising in value since there were apparently no more than three copies left unsold. The next letter from Hainhofer on 15 March probably relates to the same copy, its delivery delayed by the need to perfect it. How difficult this proved is set out in the next letter on 6 April. Anyone who has attempted to collate a copy of the *Hortus Eystettensis*, with its immensely complicated sequence of signature marks for each sheet, will sympathize with him, not to mention the 'hundred bookbinders' he mentions. Evidently, too, the problem of perfecting the copy had exceeded Besler's resources or at least the stock he had available for sale; the reference to 'the authorities' suggests that Hainhofer had to call in the Council or Guild to extract the necessary sheets from others who held stock of the book.

The price has risen to 48 fl., almost half as much again as he had originally paid; there are only two copies left, and this, if he had not bought it, would have been bought by 'Dr Mathiol' for 'Dr Faber' in Rome.[12] If Besler (and it is clear that he remained in charge of sales, of plain as well as coloured copies) has succeeded in virtually selling out 300 or more copies of such a substantial book in barely four years, the whole operation must be accounted an outstanding success. It is difficult to give a sense of relative values, but, set against a price range of 35–48 fl. for a plain copy or 500 fl. for a coloured one, the annual wages of 60 fl. that Bishop Johann Conrad paid his two head gardeners[13] or 1350 fl. paid by Besler in 1601 for a house in Sieben Zeilern, give some indication. In 1616, at the height of the success of the *Hortus Eystettensis*, he bought another house for 2500 fl., one of the grandest in Nürnberg, opposite the Lorenzkirche, where his son Michael set up the 'Mohrenapotheke', following in his father's footsteps.[14]

If, as stated, the accounts at Eichstätt showed total payments of 17,920 fl., the *Hortus Eystettensis* had indeed cost the two Bishops or the Chapter an enormous sum, but this again has to be set against the 20,000 fl. or even 50,000 fl. that Bishop Johann Conrad had been prepared to lend to the princely neighbours with whom he hunted. The ninth clause of the 'Extractus' evidently indicates an expectation that Besler would return the copper plates, presumably in 1617, and provide accounts of copies sold and unsold (this last suggests how out of touch they were at Eichstätt with what was going on at Nürnberg) and — evidently a matter of special concern — the 'dedication' (coloured?) copies.

---

12. R. Gobiet, *Die Briefwechsel zwischen Philipp Hainhofer und Herzog August d. J. von Braunschweig-Lüneburg* (Munich, 1984), no. 304 (see Appendix A, p.66). 'Dr Mathiol' is Ferdinandus, the son of Pier Andrea Mattioli, working then as a doctor in Augsburg; 'Dr Faber' is Johannes Faber (b. Bamberg, *c*. 1570 – d. Rome, *c*. 1650), botanist and physician to Popes Urban VIII and Paul V (see *Nouvelle Biographie Generale*, XVI, 896).
13. *H.E.*, 54.
14. Stadtarchiv Nürnberg, *Libri litterarum*, CXIV, ff 102–4, and CXXVIII, f4; *H.E.*, 100 and 102.

Here again, we have no knowledge of the other side of the story, Besler's. It is probable that the total costs, the payments to artists and engravers, for the letterpress printing, the rolling press for the plates, and for the two or more separate stocks of paper (binding would remain a separate charge to the buyer), far exceeded the sum outstanding in the Eichstätt accounts. Bishop Johann Conrad had certainly indicated that he expected no other return than the dedication, which Besler had been punctual to extend to his successor after his death. He may have been aware of other debts, but they were not of a monetary kind; all the records suggest that, after ten years, he could look back on the gigantic enterprise on which the Bishop and he had embarked with every ground for satisfaction.

Later history

HAD Besler felt a debt, long after the main business of the *Hortus Eystettensis* had been brought to a successful conclusion, it was, perhaps, not to the two bishops who he had served better than they knew but to his colleagues and helpers. It may be that he had not been sufficiently generous in acknowledging the help he had received. The colleagues of Ludwig Jungermann at the University of Altdorf later certainly believed that he had been largely responsible for the text of the *Hortus Eystettensis*, if not for the errors in it. Johann Jakob Baier (1677–1735) in his lives of the professors of medicine at Altdorf, published in 1728, quotes contemporary documents to this effect, even going so far as to allege that Besler was so ignorant of Latin that he had to get his brother Hieronymus to draft the preface for him.[1] None of these documents is now extant but a letter exists, dated 12 December 1614, to Caspar Hofmann from Jungermann, then at Giessen, in which he predicts that Hofmann will be unlikely to extract the seeds he wants from Besler: 'I asked as my habit is', he wrote, 'and was refused — this is the sort of gratitude I get from an overweening man whose faults speak for his character'. Parenthetically, he asks if 'opus illud grandius', clearly the *Hortus Eystettensis*, to whose title Besler has wished to add his name, is yet on sale.[2] The reference to the title-page is interesting, since it confirms what the engraving suggests, that Besler's name on it was an after-thought, a necessity, perhaps, of the new role imposed on him by the Bishop's death.

Whether Besler was guilty of plagiarism, or felt that he could use what he had paid for without further acknowledgement, cannot now be ascertained. But some sense of a debt unpaid may explain the appearance in 1627 of the *Hortus Eystettensis*, or rather portions of it, in a new form.[3] Only two (or three if the lost Erlangen copy should be discovered) copies of this exist. The title, now enclosed within the Beslerian or pharmacopoeic frame used elsewhere for the part-titles, echoes that of the original work, but without mentioning Eichstätt or the garden at all. 'Icones sive: repraesentatio viva, florum et herba[r]um: opera Basilii Besleri philiatri et pharmacopoei Norici In gratiam, Herbarum cultorum noviter, accuratâ diligentiâ, promulgata', portraits or a lifelike representation of flowers and plants: the work of Basilius Besler, lover of medicine and apothecary of Nürnberg, published in gratitude anew by the accurate care of plant-growers, the title reads, with a hexameter line in praise of horticulture which is a chronogram for the year 1627. This is followed by a page of tribute to the Collegium Medicum of Nürnberg, past and present, 'Sacri Asclepiadarum collegii, dominis doctoribus', whose names, headed 'Patronis, Cognatis, Amicis',

---

1. J. J. Baier, *Biographiae professorum medicinae, qui in Academia Altorfina unquam vixerunt* (1728), 82–3; *H.E.*, 97–8. Hieronymus Besler (1566–1632) had studied in Padua and compiled 'Collectanea optimorum medicamentorum chymicorum et aliorum', which still survives in manuscript (University Library, Erlangen, MS. 1150), with his Padua lecture notes (MS. 981) and a copy of a commentary on Aristotle (MS. 493) by Giordano Bruno, whose secretary he was in Helmstedt and Padua (*H.E.*, cat. nos. 12–13).

2. University Library, Erlangen, Briefsammlung Trew, Jungermann Nr. 2; *H.E.* 98 and n.35 (115–16).

3. Baier, 276, attributes this (which he calls 'die 3. Auflage') to the initiative of Hieronymus Besler, but without giving any evidence; there is nothing to suggest this in the text. He may not have seen the copy at Darmstadt on which his description is based.

follow. They include, in the first category, Joachim Camerarius, dead thirty years ago, his brother Hieronymus in the second, and, in the third, Caspar Hofmann and Ludwig Jungermann.

Beneath their names follows the message: 'to these who surmise what I owe in gratitude and honour, Basilius Besler, apothecary and lover of medicine, gives this little work of his, in place of requital and a monument of his gratitude.' On the opposite side of the sheet is an ornamental address, recording the inception of the work sixteen years earlier (that is, in 1611), and hoping that the accurate depiction of plants and all their appurtenances will benefit his colleagues, to whom he wishes all health.

This sounds as if it means what it says. The somewhat clumsy Latin does not read as if it had been composed by someone else for Besler, whose brother is indeed one of the beneficiaries. But what exactly is the 'little work' that is thus given away? It is clearly not the *Hortus Eystettensis* itself, as hitherto published. The two accessible copies contain only a handful of plates from it each, one with 23 plates, the other 96, together with the print of the 'maye krueg' by Georg Gärtner which Hainhofer mentioned (see p.15 above) but is not properly part of the book.[4] Perhaps the difficulties that Hainhofer experienced in perfecting the third copy he sent to Herzog August may throw some light on this. There is always a problem with the printing of any book, especially a large one like this, most of whose sheets have gone through two different printing processes: the final number of copies of each sheet varies, but no more complete copies of the whole book can be produced than of the sheet of which the least number of perfect copies has been printed. This final number may be greater or less than the desired number, depending on the allowance of extra paper made for waste during printing, but there will always be 'overs' — extra copies of some sheets, unusable because of a shortage in another sheet. Perhaps Besler made a virtue of necessity, and made the *amende* that he (or they) may have felt was due to the colleagues who had helped him by presenting them with some otherwise unwanted sheets, with a special title-page and dedication.

It was, however, still possible to acquire a copy of the complete book, notwithstanding the difficulties of which Hainhofer made so much. Besler's nephew, Michael Rupert, a medical student at Padua in 1628, wrote to his father Hieronymus on 12 March to say that he had just learnt that books could be imported from Germany, and asking for a copy of the *Hortus Eystettensis* to be sent to him with some seeds already requested.[5] It took some time to come, but it had arrived when he wrote again on 26 November 1629, although it was still inaccessible having been put in quarantine for three weeks by the Venetian authorities for fear of the plague.[6] Besler himself had died the previous March, and brother and nephew may have profited from the posthumous discovery of stock still unsold, for there is a receipt for the freight costs of a copy that they (or perhaps the whole family) gave to the University of Altdorf in 1630/1.[7]

It was probably this event that prompted the Chapter at Eichstätt to summarize their claims and try — once more — to regain their due. It seems likely that, so far as the plates were concerned, they were successful, although whether all the plates were actually transferred to Eichstätt may be doubted. They would have made a heavy load, and may have been better kept at Nürnberg, where they would be near specialist printers and a rolling-press, although both might have been imported to Eichstätt, if occasion demanded. But it did not. The invasion of the Swedish troops under Herzog Bernhard von Weimar in 1633–4 left three-quarters of the town in ruins, and it was not until Bishop Marquard II Schenk von Castell began his long reign (1638–1685) that the town and its surroundings began to recover.[8]

One of his first endeavours was the reprinting of the *Hortus Eystettensis*. For this purpose the titlepage was altered. Where before it had read that the contents were to be seen in the episcopal gardens 'hoc tempore', now (rather sadly) the word 'olim' (formerly) was substituted. Where too the word 'opera' had been followed, rather clumsily, with Besler's name, it simply read 'curis secundis'. Whose 'second care' was involved is not made clear until the

---

4. Darmstadt, Hessische Landes- und Hochschulbibliothek, and the copy formerly in the De Belder Collection (Sotherby's, 27 April 1987, lot 26), now in the same collection as the coloured copy (DeB) from the same source (lot 23).
5. University Library, Erlangen, Briefsammlung Trew: Besler, Michael Rupert, Nr. 41 (*H.E.*, 104, 117 n 57 and cat. no. 18).
6. *Ibid.*, Nr. 44 (*H.E.*, 104–5, 117 n 58).
7. *H.E.*, 117, 123, 144 n 21.
8. *H.E.*, 56–7.

preliminaries, which consist of two leaves, one headed 'Primis auspiciis' with the 1606 portrait of Johann Conrad von Gemmingen, the other, 'Curis secundis', with a similar portrait of Bishop Marquard II; the author of these pages signs himself 'Carolus Berttius'. He thanks Bishop Marquard who has made the printing of this edition possible, and put it within reach of those who desire it 'quales fuere in Gallia, Belgica, Britannia complures' — interesting evidence that the first edition had not circulated as widely as hoped, whether because local demand had been so great or because war had prevented it.[9] Only the plates were printed, with no text, perhaps due to its inaccuracy, but more likely to the exigencies of time and place. The plates show sign of wear and have not been very well printed; too often they have not been wiped carefully enough before being put to the press, so that a grey background tint has built up. Perhaps printer and press were brought to Eichstätt, after all. The paper is not unlike that used for Besler's plain or ordinary issue, white in colour, with chain-lines 26–28 mm apart; a serpent or crown watermark occurs occasionally.

The appearance of this edition and the disappearance of Besler's name did not mean that the family interest in the book disappeared. Georg Christian Stöberlin, writing from Verona to his fellow-countryman Johann Martin Brendel, from Nürnberg but then at Padua, on 20 February 1653, asks on behalf of the Italian author Francesco Pona whether he will send to Besler (Michael Rupert) the catalogue of Pona's works, enclosed, and whether a copy of the *Hortus Eystettensis* is to be had for money.[10] Michael Rupert's continued interest in the book is attested by his appendix, *Mantissa ad viretum stirpium, fruticum et plantarum . . . eystettense admirandum celeberrimum Beslerianum*, of which the dedication manuscript (for it was not printed at this time) to the Austrian Archduke Leopold Wilhelm is dated 1648.[11]

To all appearances, Eichstätt, in the person of its new Bishop, and the Besler family had come to an agreeable understanding, previous differences all settled. Indeed, the understanding was probably closer than appearances suggest. It is not only the letters cited above that indicate that copies of the *Hortus Eystettensis* could still be had from the Besler family. It is clear from study of the surviving coloured copies that one sheet that was not in short supply was the original 1613 title-page, and copies exist, as will be seen, which were certainly still available as late as 1680. Whether the Besler family still retained this stock or whether they came from Eichstätt is not certain, but the former seems the more likely alternative. Other coloured copies dating from the mid-century consist of sheets of the original 'plain' edition with the letterpress text on the verso of the plates, clearly still available then, if perhaps second-hand. What must have been one of the last copies of the original 'special' edition was also available to be coloured and sold to the Imperial Library at Vienna as late as 1678.[12] It is, perhaps, significant that the earliest date in the book, 1671, is the same year that Michael Rupert's son, Joachim Hieronymus Besler, died untimely as a medical student at the university of Altdorf, the last of the family.[13] It may be that there was a final clearance then, and that what was indeed the last surviving stock ultimately reached the market at that time.

By the beginning of the eighteenth century, all this was history, part of it somewhat garbled in the telling, as we have seen. The preservation, not only of the *Hortus Eystettensis* itself, but of its history, too, during the eighteenth century was due to three men, Bishop Johann Anton I Knebel von Katzenellenbogen, who conceived the idea of a 'centenary' edition, Johann Georg Starckmann or, as he preferred the Greek form, Sthenander (1701–80), an Eichstätt physician, who finally brought it to fulfilment, and Christoph Jakob Trew (1695–1769), physician at Nürnberg, but also one of the leading botanists of his time with a pioneering interest in the history of botany, a prolific publisher of his own and other great works of natural history in translation, whose collections, bequeathed to the University of Altdorf and, on its closure in 1809, transferred to the University of Erlangen, have been and will remain of unique importance for the study of the *Hortus Eystettensis*, its antecedents and subsequent influence.

9. The earliest copy to reach an English library that I know was, however, of the first edition, and is still at Chetham's Library, Manchester. It was bought by the Feoffees from their usual London bookseller, Robert Littlebury, on 17 June 1669 for £10, the highest price (by a long way) that they paid for a single volume in the 17th century. It was kept separately and recorded in subsequent inventories as 'The Great Herbal'. I owe this information to the Librarian, Dr Michael Powell.

10. University Library, Erlangen, Briefsammlung Trew, Stöberlin, Georg Christian, Nr 1 (*H.E.* 105 and 117 n 60).
11. *Ibid.*, MS. 2727 (*H.E.*, cat. no. 16).
12. See below (p.57).
13. *H.E.*, 101, 105.

Thanks to the correspondence of Sthenander and Trew, preserved by the latter, it is possible to study the history of the eighteenth-century edition of the *Hortus Eystettensis* in considerable detail. It was Bishop Johann Anton I's original plan, as it had been Michael Rupert Besler's, to augment the original work by adding to it. He was encouraged to do this by Johann Michael Hertel, professor of medicine at the University of Ingolstadt, who sent him drawings to be engraved. This enterprise got so far that the original titlepage was re-engraved a third time, this time with the words 'curis secundis' beaten out, and a new legend substituted: 'Cui accessit complurium florum et plantarum tum ex remotis terrae partibus quam praecipuis Europae hortis advectarum typus librum hunc denuo magis excolens curis', 'to which is added the likeness of many flowers and plants from the furthest parts of the world as from the most distinguished gardens of Europe, improving the book anew, by the care of', and the Bishop's full style follows, with the date 1712.[14] This grandiose scheme did not materialize, for reasons not made explicit at the time (it may simply have been the difficulty of getting the new figures adequately engraved). Instead, the plan reverted to a simple reprint.

Some light is thrown on this by the Diocesan archives. On 8 January 1712 the Bishop informed the Chapter that he proposed to meet the costs of paper and printing, and to provide 50 copies for the Chapter library, plus one for each member of the Chapter. As an alternative he proposed to place 'one and another copy' in the library and to offer each member of the Chapter a badge set with diamonds. The Chapter, not surprisingly, accepted this proposal, on the grounds that the library would not need so many copies. The Bishop wrote again on 12 February, that though the badges were not quite ready, the book lacked 'frontispiece and portrait', which would not be ready till next year. The Chapter replied that, since the book would be delayed, they would rather press for early delivery of their badges.[15]

But further problems emerged; two plates were missing; the other plates were no doubt worn and hard to print. The reprinting of the text, too, can have been no light or brief task for the local printers, the Hofdruckerei Strauss; if the major part was printed, the completion of the project rested in suspense. It was rescued from this by the energy of Sthenander. The missing plates were found and printed together with a revised title and a tactful introduction by Sthenander, praising the successive Bishops (and denigrating Besler), especially Bishop Johann Anton I, whose name is given in that form on the title-page, indicating that it must have been finally printed during the reign of Johann Anton II von Freyberg (1736–57). The date printed, 1713, is presumably the true date of the inception of the enterprise in the form finally agreed. The reasons for the delay and his own part in bringing it to an end are set out by Sthenander in letters to Trew in 1746; presumably the completion of this last edition of the *Hortus Eystettensis* followed soon after. There seems no reason to doubt that the letterpress was on this occasion printed at Eichstätt.[16]

Sales seem to have been slow, and there were still many copies left at the end of the reign of the last Prince-Bishop, Johann Anton III von Zehmen. In the confused events of the first decade of the nineteenth century, Eichstätt passed under the control of the (Austrian) Grand-Duchy of Tuscany in 1803 before becoming a dependency of Augsburg in 1817. The *Hortus Eystettensis* was three times offered for sale, in 1803, 1805 and 1807, at prices which, as the last advertisement in the *Eichstätter Intelligenzblatt* put it, would hardly cover the cost of the paper. During these attempts, and in an effort to promote a now old-fashioned book in the new world of the Linnean system of plant-classification, another local physician, Franz Seraph Widnmann, produced his *Catalogus systematicus secundum Linnaei systema vegetabilium adornati arborum, fruticum et plantarum celeberrimi Horti Eystettensis*, published at Nürnberg in 1805, with a French edition, dedicated to the Empress Josephine, issued from Eichstätt the following year. This was in effect a cross-reference index between the *Hortus Eystettensis* and the Linnaean system. But there were still many copies left in 1817; 78 went to Augsburg, while 100 remained at Eichstätt to be sold off in 1820.[17]

---

14. Proofs of this title page exist in the Eichstätt Diözesanarchiv and at the University Library, Erlangen (Trew B4). The latter is extensively annotated by Christoph Jakob Trew, from which the facts above are drawn.
15. Staatsarchiv Nürnberg, Bestand Hochstift Eichstätt, Protokolle des Domkapitels (Rep. 190a), Nr. 1132 (*H.E.* 106–7, 117 n 65).
16. The work must have been finished and the edition published about 1750 (there is a reference in the preface to Bishop Johann Anton II's ordination jubilee, which fell in 1749). The history of it up to this point is remarkably well documented in Sthenander's letters to Trew (University Library, Erlangen, Briefsammlung Trew, Starckmann, J. G., Nr 10 & *passim*). Trew did not acquire his own copy until 1752. (See *H.E.*, 110–11, 131–4.)
17. The sad end of the enterprise is charted by Baier, 278, and *H.E.*, 112.

It remains to chart the fate of the long-lived but by now worn-out copper plates. According to a contemporary witness, they were removed from the Willibaldsburg by the French in 1796 and taken to Neuburg, but returned when it was discovered that they were 'learned material'.[18] They remained to be transported to Augsburg with the 78 copies in 1817. Thence they were sent to the Royal Mint at Munich, and there melted down. With that, the history of the printing and publication of the *Hortus Eystettensis* came, after two centuries, to an end.

18. Hofkammerrat Joseph Barth (1760–1819), who contributed the article on Eichstätt cathedral to Johann Kaspar, *Geographisches, Statistisch–Topographisches Lexicon von Franken* (1799–1804), noted in his interleaved copy of the book (now in the Eichstätt Diözesanarchiv) these facts, with a lament that it was now impossible to discern where in the previous century one of the most famous botanic gardens in Germany had been founded, which had no equal in Europe for rare and costly plants. (*H.E.* 59–60, 67 n 159).

CREATING THE PLATES

The sources of the plates

CAN nature change? Every age sees nature with its own eyes: yet that vision is complicated and varied by what previous ages have seen and how they have recorded what they regarded as the essential features. In the middle ages accuracy, in the sense of an objective record of what was seen, was not so important as the faithful copying of an existing depiction. Nature was not to be questioned but followed; the essential quality of the thing depicted lay in a previous model, sanctified by an implicit connection with the Creation itself and thus to be passed on without question, but inevitably varied by the unconscious artistic conventions of the time. It is not until the end of the fourteenth century that the new humanism in Italy and the humanistic habit of thought produce a new, objective view of nature. The famous Carrara Herbal (B.L. Egerton MS 2020), made for Francesco da Carrara, the last Carrara ruler of Ferrara (d.1403), is remarkable not only for the accuracy of its pictures, but also for their independence. The plants depicted are recognizable to us today, and answer our vision, influenced by photography, of what the 'real' plants look like. At the same time, the pictures break out of the conventional rectangle of the written page and struggle to escape from the page itself. Complete independence was not possible: you cannot paint a tree life-size, nor yet every leaf on it. A compromise had to be reached, not only with practical necessities, scale, vehicle, medium, but also, again, with the artistic conventions of the time.

Striking and unconventional as the *Hortus Eystettensis* was in terms of content and the ordinary norms of book-production, it too represented a compromise, and one that can only be understood by examining its more remote, as well as immediate, antecedents. The norm for the printed depiction of plants was set by the herbals of the early part of the sixteenth century, in particular the much reprinted works of Otto Brunfels, Hieronymus Bock and Leonhart Fuchs.[1] These were illustrated by woodcut blocks, made up with the type of the text on the page. In this they resembled the earliest herbals, like the ever-popular 'Hortus Sanitatis', plain copies of medieval images, but the image reflected the new humanistic sense of realism. Each block depicted one plant, with varying degrees of accuracy, depending on its habit and use (the latter might demand flower and fruit to be shown simultaneously). The relatively narrow dimensions of the common small folio page made a vertical oblong the norm for the woodcut blocks as well.

Some of these early books are cited in the list of 'Auctores, qui in opere Horti Eystettensis citantur', which begins, indeed, with Pliny.[2] Such works are, however, the sources for the text, not the plates. These must be sought elsewhere, among the pioneers of the new study of botany for its own sake, rather than for medicinal purposes. Accurate depictions of plants, particularly those newly discovered (tulips from Turkey, as well as more exotic plants from the Americas and Far East), were essential to this new scientific purpose. The great pioneer, Conrad Gesner of Basel (1516–65), made and collected drawings of flora and fauna, but lived to publish only the latter. His drawings, however, were used by others during his life-time (the interrelation of collections of drawings with the more generally accessible illustrated printed books remains to be studied). They have a particular importance since they were acquired after Gesner's death by Joachim Camerarius the Younger.[3]

---

1. Otto Brunfels (1488–1534) *Herbarum vivae icones* (Strassburg, 1530), Hieronymus Bock (1498–1554) *New Kreütter Buch* (first illustrated edition, Strassburg, 1546) and Leonhart Fuchs (1501–66) *De historia stirpium* (Basel, 1542).
2. A total of 35 different books by 21 authors or editors are cited, ranging in date from 1543 (Fuchs) to 1611 (Clusius, *Curae posteriores*). The more significant are discussed below.
3. The drawings, on 490 leaves and bound in two large volumes, passed from Camerarius to his son Ludwig Joachim (1566–1642), from him to Johann Georg Volckamer (1616–93), intimate friend of Ludwig Jungermann, who left them to his son, also Johann Georg (1662–1744), thence to Christoph Jakob Trew, with whose collection they followed the now familiar route to the University of Erlangen (MS. 2386). They have recently been reproduced in facsimile, *Conradi Gesneri Historia Plantarum* (Zürich, 1972–80).

The influence of Camerarius on the *Hortus Eystettensis*, if undocumented, is so strong that it is worth pausing to reflect on his life and work.[4] He was born at Nürnberg on 6 November 1534, and was educated first at home and then at the school where his brother-in-law Esrom Rüdiger taught. He was a particular favourite of the great Philip Melanchthon, who treated him like a son. At first, he followed the example of his famous father, as a classical scholar; then he turned to medicine, which he studied first at Wittenberg, then at Leipzig. He was two years at Breslau, studying under the famous doctor Johannes Crato, a friend of his father's. He then went to Italy, spending a year at Padua, and then a further year at Bologna, where he received his doctorate on 27 July 1562. He returned to Nürnberg, becoming 'Stadt-physicus' to the city in 1564. There he continued to live, founding the Collegium Medicum, of which he was the first president, in 1592. Most of his time was devoted to his botanical studies, establishing his own botanical garden ('einen prächtigen Garten') outside the city; he also set up a similar garden for the Landgraf Wilhelm of Hesse, at Cassel. He died on 11 October 1598.

The garden that Camerarius kept had in fact previously belonged to Georg Oelinger (1487–1557), 'Materialist' and apothecary of Nürnberg, where he had premises 'am Krebsstock' by the Fleischbrücke, and of which he was a member of the 'Grösser Rat'.[5] Camerarius also acquired the remarkable collection of natural history drawings, mainly by Oelinger himself, with some assistance from Samuel Quicchelberg, later keeper of the botanical collections of Duke Albrecht V of Bavaria, and other hands. The possession of this collection, and that of Gesner, make it clear that the depiction of plants, as well as their cultivation, was a matter of primary interest to Camerarius. Some at least of the Gesner drawings appear to have provided the new illustrations ('vielen schönen newen Figuren') for the *Kreutterbuch*, the German translation of Pier Andrea Mattioli's *Historia plantarum* (1561), which he produced and Feyerabend printed at Frankfurt in 1586.[6] This and his other main botanical work, *Hortus medicus et philosophicus* (Frankfurt, 1588), are much cited in the *Hortus Eystettensis*.

References in these works make clear the extent of his scholarly correspondence. In Italy, besides Mattioli, he knew the great doctors, Fallopio and Fabrizzi, Prospero Alpini, who travelled to Egypt collecting specimens and later became keeper of the botanic garden at Padua, and Ulisse Aldrovandi, founder of the famous garden at Bologna; he also knew Jacopo Antonio Cortusi, who succeeded Alpini as director of the garden at Padua and like him went on botanical journeys in the Greek islands and the Near East, Jacques Dalechamps, the botanist at Lyon, and Joost Goedenhuyze, the Flemish botanist who became keeper of Francesco de' Medici's botanic gardens at Florence and changed his name to Giuseppe Casabona.

But of all his foreign acquaintance, the closest was that other more famous Flemish botanist, Carolus Clusius (Charles de l'Ecluse, 1526–1609). They had met as students of Melanchthon in Wittenberg in 1549, from which Clusius went on to Montpellier to study under Guillaume Rondelet, and thence to collect plants in Spain and Portugal for two years. In 1573 he was invited by the Emperor Maximilian II to become his personal physician and to establish and direct the botanic gardens at Vienna. There he was able to take advantage of the close diplomatic links with the Ottoman empire to import new plants from the Near East, in particular tulips. The introduction of these exotic flowers was a matter of close interest to Camerarius, and he recounts its history in *Hortus medicus*; the first tulip to be depicted was one grown at Augsburg where it was seen and drawn for Gesner (the drawing was engraved on wood and published in 1561 by Gesner in his collected edition of the works of Valerius Cordus, to which he added his own *De hortis Germaniae liber recens*), followed shortly after by Mattioli (1565) and Dodoens (1569).[7]

The succession of Rudolf II, to whom Clusius was *persona non grata* as a Protestant, brought about his exile and the closure of the gardens (he sent specimens thus released to Camerarius). He moved to Frankfurt and advised

---

4. The best account is still that of Georg Andreas Will, *Nürnbergisches Gelehrte-Lexicon* (Nürnberg, 1755), I, 173–6. Further details are in the introduction by N. Harms and V.-B. Kuechen to the facsimile of Camerarius *Symbola et Emblemata* (Graz, 1988), 2*–18*.
5. E. Lutze, *Die Bilderhandschriften der Universitätsbibliothek Erlangen* (Erlangen, 1936), 71.
6. On 28 March [1585] Joachim Jungermann (see below, p.28), Camerarius's nephew, wrote to his uncle that some of these drawings might be suitable for publication. E. Schmidt-Herrling, *Die Briefsammlung des Nürnberger Artztes Christoph Jakob Trew in der Universitätsbibliothek Erlangen* (Erlangen, 1940), 311 (no. 21).
7. V. Cordus, *Annotationes . . . de materia medica*, ed. C. Gesner, 1561, ff 213–14; R. Dodoens, *Florum, et Coronarium odoratarumque nonnullarum historia*, 2nd ed, 1569, 205, 207. See S. Segal, *Tulips portrayed: the tulip trade in Holland in the 17th century*, 1992, 3.

Wilhelm of Hesse, whose garden Camerarius had established, and then in 1587 went to Leiden where he succeeded Rembert Dodoens as professor of botany; he also re-founded the University Botanic Garden and spent the rest of his life there. All this is recorded in the 195 surviving letters from him to Camerarius, full of revealing details about the growth and acclimatization of new species. There are references to drawings, some by Camerarius's nephew, Joachim Jungermann, whose early death while plant-collecting in Crete in 1591 was a grief to both men.[8] It seems probable that Clusius, directly or indirectly, acted as adviser to the Bishop and Besler after the death of Camerarius. In his address to the reader in the *Hortus Eystettensis*, Besler cites him by name alone among the many contributors to the enterprise, and later, alluding to his own long participation (the Bishop, he says, called him 'pater et patronus' of the garden) describes it as continuing 'even after the death of Clusius'.

If Camerarius (and, through him, Clusius) was the strongest influence on the formation of the *Hortus Eystettensis*, both on the Willibaldsburg and on the pages of the great book, other names are cited. Caspar Bauhin (1560–1624) was a contemporary of Camerarius's nephews, and published his *Phytopinax* in 1609, followed in 1623 by *Pinax theatri botanici*, in which plants were classified by genus and species; it was to have a formative influence on Linnaeus, the more so since it provided a concordance of earlier plant names. Castore Durante (1529–1590), personal physician to Pope Sixtus V, published his *Herbario nuovo* at Rome in 1585; this was translated into German by Peter Uffenbach and published at Frankfurt in 1609 under the title of *Hortulus sanitatis*. One of the earliest local floras, Johannes Thal's *Sylva Hercina*, covering the Harz mountains, was published under the auspices of Camerarius and is usually found bound with *Hortus medicus*. A later work also cited in the *Hortus Eystettensis*, Caspar Schwenckfelt's *Stirpium et fossilium silesiae catalogus* (1600), covered a wider area, including some cultivated species. The more traditional, and popular, *Neuw Kreuterbuch* of Jakob Theodor of Bergzaben, hence 'Tabernae-montanus', is also among the works cited; first published in 1588, it was later edited by Bauhin.

Perhaps the most enterprising and novel work, by contrast, was Leonhart Rauwolff's *Aigentliche Beschreibung der Raiss . . . in die Morgenländer* (Lauingen, 1583). The author, famous in his own lifetime for his plant- and seed-collecting journey to the Near East, from which he brought back over 800 species, is commemorated today by the remarkable herbarium drawn from Southern France and Italy, now preserved at Leiden. In his lifetime, he practised medicine and kept a garden at Augsburg, where he corresponded with Clusius and Gesner. Driven thence by religious faction, he entered Austrian service as a military physician, and died at Hatvan in Hungary in 1596.

The work of Matthias de l'Obel forms another significant factor in the origins of the garden of Eichstätt, and one of an entirely different kind. De l'Obel himself (1538–1616) was another botanical pioneer. Like Clusius and Bauhin he had studied medicine at Montpellier, and practised as a physician in Antwerp and Delft before coming to England in 1569; he returned to the Netherlands in 1571 but settled permanently in England in 1585, and was appointed royal botanist by James I. He was the author of *Plantarum seu stirpium historia* (1576) and *Plantarum seu stirpium icones* (1581), both printed by Plantin at Antwerp. The first is interesting because De l'Obel tried to arrange the plants in an order based on structural similarity, thus in some cases anticipating the Linnaean classification. The second work, though smaller in bulk, was equally original. It was designed in an oblong format, with a few lines of text and a woodcut picture for each plant. The number of plants in the *Historia*, 1441, was increased to 2173. The woodcuts, though simple in design (mainly outline drawings, with minimal shading) are extremely 'accurate' — that is, they have the objectivity necessary in a good recognition manual.

It may have been this, or the convenient amount of white paper which the format afforded, that recommended the book for annotation and expansion to Johann V Molitor (Müller, *d.*1627), Abbot of the Benedictine Abbey of Michelsberg, Bamberg.[9] He was of a scientific disposition, interested in mathematics, in clock-making and in illuminating manuscripts and printed books, and also, on the evidence of this book, in botany. As well as adding drawings and notes, often with the dates and locations, in the gardens of contemporaries in Bamberg, of the flowering

---

8. F. W. T. Hunger, *Charles de l'Escluse, Carolus Clusius, Nederlandsch Kruidkundige* (The Hague, 1943), II, 295–449 *passim*.
9. I am much indebted, in what follows, to the discussion of Abbot Johann's works in Brigette Hoppe, 'Kräuterbücher, Gartenkultur und sakrale decorative Pflanzenmalerei zu Beginn des 17. Jahrhundert', *Rechenpfennige: Aufsätze zur Wissenschaftsgeschichte Kurt Vogel zum 80. Geburtstag . . . gewidmet* (München, 1968), 183–216.

or habit of individual species, he also used his copy of *Icones stirpium* as a *hortus siccus*, and from time to time added a faded leaf, notably of *Amaranthus tricolor*, a recent discovery from tropical Asia, whose brightly coloured, almost metallic, leaves had given it the name 'Papageyfeder', parrot's feather.

In his notes he cites other books, many of them also quoted in the *Hortus Eystettensis*, besides De l'Obel. There are tantalizing references to three other books, similarly 'illuminirt', by Mattioli, Tabernaemontanus and Clusius. The author most cited of all is Camerarius, notably his *Hortus medicus*, abbreviated to a large capital C with a stroke through it, followed by 'in horto'.

All this apparatus can, in turn, be linked to the spectacular decoration of the vaults of the nave, the north and south aisles and the transepts of the Michaelskirche, painted as part of the restoration of the church during Abbot Johann V's abbacy in 1610–14. There are about 600 different plants and trees depicted in the vaults, many of which can be identified as copied from woodcuts in Fuchs, Bock and Lonitzer, as well as De l'Obel. In several cases, however, the ceiling paintings can be shown to derive exactly not from woodcuts, which would be common to any copy of these books, but from Abbot Johann's extra-illustrated copy, either following adaptations of the woodcuts or, in two cases, wholly new additions. The fine example of *Crocus biflorus*, entirely drawn and noted as flowering in March 1615, appears in exactly the same form in the nave of the church.

Startlingly original though this form of decoration is, especially on this scale, it is not unique. The great church of St Kilian, Würzburg, was similarly painted with flowers in 1608. The name of the artist is known, Andreas Herneisen from Nürnberg, but the pictures no longer exist, since the church was redecorated in baroque style in 1701–4, and the only visual relic is a painting of the interior of the church by Barthel Büeler, dated 1627. Other churches with similar decorative plans are recorded or still exist, sometimes over-restored, at Dettelbach (1611–14), painted by Hans Ulrich Stimmer of Schaffhausen and Würzburg, Rothenfels (1611–12), Karlstadt am Main (1614) and Buchold (1622), the last the work of another Nürnberg artist, Wolfgang Ritterlein.[10] Perhaps the closest, in terms of systematic design, to the Michaelskirche at Bamberg was that at the abbey church at Ebrach, executed in 1614 but substantially diminished by the severe damage to the church in 1631 during the Thirty Years War, and its subsequent restoration in baroque style. Even nearer, geographically, was the former bishop's residence at Bamberg, Schloss Geyersworth, which also had botanical decoration dating from 1593, although by this time it had been abandoned in favour of Seehof by Bishop Schönborn.

If, then, we can detect in and around Franconia and upper Bavaria a desire, if not common, at least widespread, for decorative church painting based on contemporary, and life-like, botanical illustration, what did it signify? The central vault, at the intersection of the transepts, at the Michaelskirche, provides the beginning of an answer. The very centre of the vault has four pictures, arranged in a cross-shape — the olive, the laurel, the bitter orange and the cherry (*Olea europaea*, *Laurus nobilis*, *Citrus aurantium* and *Prunus cerasifera*); while around the edges are the date-palm (*Phoenix dactylifera*), the vine (*Vitis vinifera*), the white mulberry (*Morus alba*), the sweet cherry (*Prunus avium*), the orange (*Citrus aurantium*) or 'Adam's apple' as it is labelled in Abbot Johann's neat manuscript, the peach and apricot (*Prunus persica* and *P. armeniaca*), the Indian fig (*Opuntia ficus-indica*), a substitute for the ordinary fig, standing for peace, and its opposite, the apple (*Malus communis*), the bladder senna (*Colutea arborescens*, locally known as 'Palmzweig', palm branch), and, finally, the passion-flower and holly (*Passiflora caerulea* and *Ilex aquifolium*). It is not hard to see in this the figure of Christ in the plant depicted in the centre, while, round the edge, are those that relate the events of his life and the Old Testament 'parallels' that prefigured it.[11]

Spreading out in all directions from this central message are all the plants in creation, at once typifying the earthly and the heavenly paradise; in this, the newly discovered plants from distant lands have a special importance as an extension of the connection between mankind and nature, an enlargement of the symbolic means by which the works of God can be justified and explained. It is perhaps a mannerist conceit in the age of mannerism, but also one deeply rooted in an older Christian tradition.

10. Büeler, *fl.* 1630 at Nürnberg, Stimmer, 1589–*post* 1625, and Ritterlein, *d.* 1622, are mainly known for the works described (Thieme-Becker, *Lexicon der bildenden Künstler*, V, 189, XXXII, 57 and XXVIII, 390).

11. L. Behling, *Die Pflanze in der mittelalterlichen Tafelmalerei* (Weimar, 1957), 44–8, and *Die Pflanzenwelt der mittelalterlichen Kathedralen* (Köln/Graz, 1964), 81f, 96–9, 103f; Hoppe, *loc. cit*, 195–6.

What point of contact, if any, exists between Bishop Johann Conrad at Eichstätt and Abbot Johann V Molitor at Bamberg? At the simplest level, none: Eichstätt and Bamberg, so near now, were then several days apart, and there is no reason to suppose that the protagonists in either place ever met or that there was any direct influence in either direction. At the same time, a common outlook, as well as a common source of contemporary reference, was familiar to both: both attempted to acquire out of the way plants and to acclimatize them in very similar surroundings. Beyond this, again, was a common attitude to the real end of botany, a desire to glorify God through better knowledge of His creation. But at this point, a considerable variance can be detected. The Abbot took nature as it came, preferring the simplest, most 'life-like', representations, and by setting them in a sacred context allowed them to speak for themselves. The Bishop, first choosing the plants for his garden, then arranging some in tubs and pots, and arranging these with others planted in the eight separate gardens, imposed his own order on the garden, an order which (we may suppose) was expressed again in the book, partly in its seasonal arrangement, but more pervasively and subtly in the arrangement of different plants on the page. If less directly exegetic than the Abbot, the Bishop was at equal pains to display the full breadth of natural creation, to fulfil, so to speak, the promise of the Garden of Eden offered by the title.

The pages of the *Hortus Eystettensis* are less realistic than the woodcuts that recur in the books by De l'Obel and Dodoens printed by Plantin. Whereas each woodcut block is a separate unit, one for each plant, the great copper plates of the *Hortus Eystettensis*, whether they depict part only of a single plant or several complete small ones, are composed. The artistry of these compositions is no doubt itself composite, the product of broad direction by the Bishop at one end of the scale and many small decisions by the various artists, from the first draughtsman to the engraver, at the other. At some point, an impresario or layout artist, by selecting what was to go or could be fitted on each page, exercised an influence both on the overall content and the individual depiction of each plant. The change from simple *trompe l'oeil* to the subtler conspiracy to delude sense by placing its objects within the larger pyramid of a theocentric universe is the real claim to originality, as well as the technical virtuosity of its plates in the *Hortus Eystettensis*. How this effect was achieved, and with what support, botanical and technical, must be our next concern.

Transfer from nature to plate

LETTERS from Clusius to Camerarius are full of references to the need for accurate pictures of plants. Pictures were an essential supplement, for botanical knowledge, to the seeds or bulbs by which plants might be acclimatized. Camerarius had drawings made, which he sent to Clusius for criticism. The collection of material that he came to possess, if it now included the collections of Gesner and Oelinger besides what grew in his garden, represented a substantial number of species, including many newly introduced and still unfamiliar. Gesner's drawings, in particular, showed the importance of the selection of detail, flowers, leaves, seeds, roots, bulbs, habit, in depicting the essential or recognizable quality of plants, including enlarged or separated elements. All this may well have suggested to Camerarius the desirability of a more elaborate, pictorial version of the *Hortus medicus et philosophicus*, which others, besides the Abbot of Michelsberg and the Bishop of Eichstätt, may already have found useful as a shopping list as well as a guide to simple recognition or, ultimately, to the knowledge of God.

If the notion of such a book had already been discussed, it may explain why Joachim Jungermann wrote to his uncle on 28 March 1585 approving the use of some leaves from Oelinger's 'Imagines' for publication.[1] This may also explain the appearance and form of the 'Florilegium', closely associated with Camerarius, recently re-discovered and now happily returned to the University Library at Erlangen.[2] These drawings are uniform in style and lettered in a single if somewhat variable hand; although the original layout has been disturbed when the drawings were remounted when bound in their present form in the eighteenth century, they are in much the same order. This does not conform in detail either with the *Hortus medicus* or with the *Hortus Eystettensis*, but, with some exceptions, the same broad pattern can be seen, a seasonal movement from spring to autumn, with two exotic outcrops, tulips and lilies at the beginning, nasturtium and the 'Marvel of Peru', cactus, tomato, pepper and agave, towards the end.

But it is the style of the drawings that is particularly interesting. There is an element of caricature, strongly contrasted greens for the upper and lower side of leaves, over-emphatic curves to stems and tendrils, colours even brighter and more contrasted than in nature. Some of the drawings show flowers in blue and white pots (oriental Delftware?), in boxes or on a trellis, which further emphasizes colour, size and habit. Other drawings, by contrast, are subdued in colour and design, more 'accurate'. It is as if a draughtsman, perhaps Joachim Jungermann himself who provided drawings for some of the woodcuts in the *Hortus medicus*, were consciously developing a style that would ensure that the important characteristics would survive even when transferred to the colourless medium of print. There are, in fact, not many new illustrations in the *Hortus Medicus*, but three of them, 'Aloe Americana', 'Nasturtium Indicum' and 'Ros Solis', bear a striking resemblance to their counterparts in the *Hortus Eystettensis* (352, 294 i and 312 iii).

The texts written in the 'Florilegium' are, with one exception, uniform in the style of wording; they are not notes, but are restricted to nomenclature, generally Latin and German (each in the appropriate hand), sometimes more, if the plant has not yet acquired a settled name. In one instance only further, and revealing, information is supplied. This records the double flowering of a *Pancratium*, 'Hemerocallis valentina Clusii', as it is here called, 'quae in nostro horto anno '89 bis floruit, mense Maio, & in fine Augusti'.[3] This was shortly after Jungermann left for Italy and before he set off on the expedition to Greece from which he never returned. Perhaps this sad event may explain the present form of the 'florilegium'. Everything about the book suggests an enterprise completed up to a point, but needing some finishing touch to give it the form intended for it. It may be that this would have converted it into an edition of multiple copies.

It is the nature, rather than the form, of the Camerarius 'Florilegium', that suggests a prototype of the *Hortus Eystettensis*: the seasonal arrangement (otherwise commonplace), the combination of rare and exotic plants and trees

---

1. See above, pp.24–5.
2. The 'Camerarius Florilegium' was sold by Christie's (20 May 1992, lot 151), with a long and well-illustrated catalogue by Sandra Raphael. It was purchased by the KulturStiftung der Länder, and is now in the University Library, Erlangen.
3. Florilegium, f.192.

with local and indigenous species, the restriction of the information to names alone, but, even more, the tendency to caricature — the suggestion of distinctive characteristics by over-emphasis. This was a mannerist age: part of the lasting appeal of the *Hortus Eystettensis* is a self-conscious mannerism.

But if the 'Florilegium' was a prototype, it remained such. Its further development may have been cut short by the death of Joachim Jungermann, but the details that are interesting to us may have been intended only as a matter for discussion betwen him and his uncle. The conspicuous feature of the compilation, then and now, is that it was not a private notebook, but a plan for something with a wider circulation. Camerarius had been brought up in a house where the mechanics of printing and publishing had been familiar; one of his tutors, Ernst Vögelin, had afterwards been his printer. Not merely achieving results, but publishing them, was a régime inherited from his father. At the same time, it was a very different thing to develop a project like this at home, with a remarkable collection of drawings and live specimens at hand, the more so if, as is likely, his nephew was at once coadjutor and draughtsman. To recreate this on a larger and more formed scale would be an altogether different problem.

As we have seen, there is no way of telling when Bishop Johann Conrad von Gemmingem began to plan his garden, when (if at all) he involved Camerarius in the project, and whether, if so, Camerarius inspired the Bishop to take on a project planned by his nephew and himself. As with the Michelsberg roof, there is a strong correspondence between the contents of the *Hortus medicus* and the *Hortus Eystettensis*: of the 667 plants in the latter, 462 are in the former.[4] But the precise figures are in themselves unimportant, as is the proportion; both might be matched without great difference in other contemporary works. It is the resemblance of the form and plan of the *Hortus Eystettensis* to what we know of Camerarius's objects and interests that is striking.

To reconstruct these on a vastly larger scale, over a considerable distance, and with a perceptible difference of tone, meant a change at several levels. In theoretical terms, there was the change from a plan based on a private botanic garden shared with expert friends to one based on episcopal resources, where even a personal enthusiasm became a public statement, perhaps (as suggested earlier) a theological proposition, a call for a return *ad fontes* in the wake of the Counter-Reformation. In practical terms, it involved finding artists to draw live flowers, converting those drawings into engraved plates, printing, publishing, and so on. All this might have been undertaken by Camerarius, had he lived. But he did not: his death at the end of 1598 must have left the Bishop without the support needed to carry out the plan before the garden was half built, let alone the book begun.

Bishop Johann Conrad had a great deal else to do, both in the diocese as a whole, in Eichstätt, and on the Willibaldsburg. It was, by Besler's calculation, some nine years after the death of Camerarius before work on the book was to begin.[5] By then, presumably, many of the plants acclimatized by Camerarius were flourishing on the Willibaldsburg, either indoors or in one of the eight parterres. Others would have been acquired from the merchants in the Low Countries. There need not, any more than with the 'Florilegium' of Camerarius, be an exact correspondence between the contents of the actual garden and those of the book. Indigenous plants and wild flowers could be left to flourish outside the actual garden. Others, as with Camerarius, might have to be preserved in pictorial form only — as we have seen, the possession of 'contrafetten' was in some ways more important than owning the things itself.[6] But, with all this, there would be a large residue which would have to be drawn *in situ*.

So the first step would be to engage artists competent to render botanical form and bring them to Eichstätt. While these drawings could have been monochrome (since they were destined for print), it is much more likely that they would have been coloured; there would have been little point in the Bishop sending 'boxes' of fresh flowers every week for the guidance of the artists if they were not working in colour; in the end, too, the colouring artists would need guidance. This process, dependent on the flowering times of the plants and trees, would inevitably stretch over a year, and (making allowance for failure, both botanical and artistic) probably more than one year.

It is unlikely that any other craftsmen involved in the book would have come to Eichstätt. The documents

---

4. Schwertschlager, 50.
5. See above, p.14.

6. See above, p.1 and Appendix A, p.60f.

quoted in the 'Extractus' (p.66) show that 50 plates were completed during the Bishop's lifetime at Augsburg and therefore in the workshop of Wolfgang Kilian, his father-in-law and teacher Dominicus Custos (Koster) and his son Raphael Custos, a day's journey to the south; the other engravers, Georg Gärtner, Johannes Leypold, Levin van Hulsen and his son Friedrich, Peter Isselburg, Servatius Raeven and Heinrich Ulrich, all lived and worked in Nürnberg, almost as far to the north. Their tools and equipment, the means of proofing plates, would all be to hand, and all that followed — the purchase of paper, printing, and so on — was better managed in a main urban centre. What was needed was, as we have seen, an impresario: someone with botanical knowledge, artistic sense and administrative ability to whom the Bishop could confidently delegate the implementation of the long-cherished plan. This can only have been Besler. There is no sense that he was managing two rival teams, although the plates specifically connected with him are all by Leypold. The work of each engraver seems to have been spread evenly through the book. There was a pronounced Dutch influence among the Nürnberg engravers; the Van Hulsens had come from Holland in 1590, and Isselburg and Servatius Raeven had studied under Crispin van de Passe and Jan van der Straeten (Stradanus).[7]

In fact, the whole process of translating the garden into print seems to have been very much more complicated than the simple copying of drawings on to copper. It is a process that needs to be approached from both ends. Starting from the end-product, we have the prints, 367 plates of flowers, of which 74 are signed by their respective artists (see Appendix D). The plates no longer exist, since they were melted down 170 years ago. But the drawings made for them are still extant, preserved by the admirable Christoph Jakob Trew (how they came into his hands is another matter that we shall have to consider), and now at Erlangen University Library (MS. 2370).[8] The drawings are, for the most part, monochrome, and exactly like the prints but in reverse. They cannot have been made from the life, if only because they often combine plants which could not have flowered simultaneously. But they do contain written notes on colour and in a few cases have been coloured, usually only in part. These details are less striking than the uniformity of the drawings in style and technique. They are clearly all by one hand, and one intimately involved with the engraving process; the fidelity of the engraved plates is a tribute to the artist's familiarity with the process for which he is drawing. Given this, I am inclined to wonder whether they are not the work of Wolfgang Kilian, the most famous of the engravers whose signatures or monograms are found on the plates. His name in full on the title-page may be a relic of what was originally a greater degree of participation.

Setting aside the colouring and directions for the moment, there is one other, very inconspicuous, feature of the drawings that requires explanation. A number of the drawings have *each plant* marked in faint lead point with a unique graphic sign. No two subjects have the same mark; sometimes the marks on one drawing are connected (thus, on the drawing for plate 67 the individual subjects, numbered on the plate I–V, have one, two, three, four and five rings against each drawing); more often, each mark is wholly different from the other marks on the same sheet, and has no visual or numerical connection. The marks look more like brand or mason's marks. What is the function of these marks? Obviously, they are intended to be a guide to someone not perhaps illiterate, but more used to visual than verbal instruction, who needs to know that this subject in the composite 'plate' drawing must be related to some other object, rather as signature marks in books were designed to instruct the bookbinder on the correct order of the leaves in a quire. These marks cannot be intended to correlate the drawing as a whole to the relevant plate, since drawing and plate are so like that anyone could match them; nor would there be any need to match individual subjects within the drawing to those on the plates.

The only alternative is that the marks were intended to match other drawings to the monochrome plate drawings.

---

7. The names of the engravers are given by Nissen, *Botanische Buchillustration*, I, 70–3, and brief biographies appear in Thieme-Becker, *Allgemeines Lexicon*, as follows: Wolfgang Kilian (1581–1662), XX. 302–5; Dominicus Custos (*c* 1560–1612), VIII. 219; Raphael Custos (*ante* 1590–1651), VIII. 220; Georg Gärtner d.J. (*c* 1575–1654), XIII. 44–5; Johann Leypolt, chiefly recorded at Würzburg (*fl.* 1607–19), XXIII. 174; Friedrich van Hulsen (*c* 1580–*c* 1660), XVIII. 114; Peter Isselburg (*c* 1580 – *post* 1630), XIX, 265–6; Servatius Raeven (*fl. c.* 1610), XXVII. 561. According to Doppelmayr (162–3), Levin van Hulsen came from Holland to Nürnberg about 1590 and was still active in 1606. Friedrich Ulrich appears frequently, like Gärtner and Schneider, in the Nürnberg Rat minutes, between 1597 and 1617 (Hampe, II, *passim*).
8. See below, p.49.
9. See below, pp.66–7.

Why should this be necessary, and who would benefit from it? The need must have existed because the other drawings, individual studies of plants or parts of them, were not immediately recognizable by those without botanical knowledge as pictures of the same subject as that on the 'plate' drawing. Those who would benefit from this identification system were the colourists of printed copies. If this is so, it explains why the 'plate' drawings survived — they were, in conjunction with the individual (and, we can assume, coloured) drawings of plants, an essential guide. This further explains the scanty verbal instructions on colour and the actual colouring of certain details, using rather warmer colours than were in fact generally used. These were to provide guidance to a master colourist on verbal instructions to be given to his workshop. The actual colouring is densest over the eleven sheets of tulips, subjects at once unfamiliar and extremely like one another — here, it was obviously felt, the likelihood of confusion was at its greatest. As we shall see, the provision of such precautions was not needless; even so, errors crept in.

Going back, now, to the beginning of this process, what can we infer about the early stages? First, that there was a multiplicity of drawings, perhaps by many hands, some from life, some from the cut specimens in the Bishop's boxes, others perhaps copied from other drawings, one or more of which served as 'copy' for the artist of the 'plate' drawings (Kilian, if it were he). Secondly, that some of these drawings continued to have a useful purpose after the plates were engraved as a guide to the 'illuminators' of the coloured copies of the *Hortus Eystettensis*. We have seen that the 'plate' drawings have survived, preserved for and by Trew, who was able (although this is, again, to anticipate) to acquire these relics of the colouring workshop. Is there any reason to suppose that any of the other, separate, coloured drawings have survived, since they have clearly become detached from the 'plate' drawings?

There is, in the Archives of the Royal Botanical Gardens at Kew, a manuscript *florilegium* known as the 'Kalendarium' of Sebastian Schedel, whose large painted armorial book plate, dated 1610, it bears on the front pastedown. The book still preserves the original boards covered in red morocco gilt, but it has been rebacked, probably *c.* 1900. There is no precise record of its acquisition, but it appears in the catalogue of the library published in *Kew Bulletin* in 1899; it is not listed in the first surviving account book, which begins in 1890, nor in the list of books acquired from Sir William Hooker on his death (the foundation collection) in 1865, and must have been acquired between these *termini*. The manuscript is a small folio in format: the quires of similar paper, diversely watermarked, are of uncertain length; these leaves have been foliated in a contemporary hand [1]–289; several leaves are missing and other leaves, mostly of thinner paper, have been inserted. The coloured drawings are, again, in seasonal order, with the same extensive collections of tulips, and, later, the other exotics, the 'Marvel of Peru', *Amaranthus tricolor*, *Opuntia*, and so on. Unlike the *Hortus Eystettensis*, it is not divided into four seasons, but into months. Originally there must have been up to twelve headings, of which only 'Apprilis' (f.49), 'Maÿ (f.127), 'Junius' (f.166), 'Jullius' (f.205), 'September' (f.258), 'October' (f.276) and 'November' (f.283) still exist, all written in a neat large textura.

The drawings vary in quality and detail; some are highly finished, others are merely sketches; all show a high degree of competence and botanical knowledge; but while some have the individuality of studies from the life, others are more studied and laborious, as if the artist were adapting them to a formula. The full contents, with the corresponding references to the plates of the *Hortus Eystettensis*, show how close the overall correspondence is (see Appendix D). Few of the illustrations are *exactly* like those of the same species in the *Hortus Eystettensis*: either there is a detail in place of the whole, or the image is reversed, and so on. The close parallels are marked and comparison between manuscript and printed page puts the connection beyond any doubt. Any suspicion that the manuscript might be copying the prints can be dismissed; the drawings are never quite like enough, and they show an independent familiarity with what is clearly the same botanical image. Any final doubts as to the connection between the 'Kalendarium' and the *Hortus Eystettensis* can be laid by the presence in it of a number of fragments of a coloured copy of the printed book. Altogether, eleven parts (some divided) of eight different plants from a coloured copy of the *Hortus Eystettensis* have been cut out, following the outline of the print exactly, and mounted in the 'Kalendarium', usually on versos and facing the page with a relevant drawing on it. The presence of these fragments is clearly an afterthought. Furthermore, some drawings of plants which do not appear in the *Hortus Eystettensis* have been treated in the same way; that is, they have been cut out round the outline and pasted into the book, in this case on rectos.

What are we to make of these details and, in general, the relation of this book to the printed book and, in particular, the colouring of copies of it? Before attempting to answer these questions, let us turn to the owner of the book and consider how he came to be involved, as he clearly was, with the manufacturing processes of the *Hortus Eystettensis*. The Schedels were one of the oldest and most distinguished families in Nürnberg. Our Sebastian was, in fact, the great-grandson of Hartmann Schedel (1440–1516), author of the famous 'Chronicle'. His grandfather, Sebastian Maria (1494–1541) occupied a series of important positions in the city, and his father, Melchior (1516–71), was a soldier, knighted and ennobled by the Emperor Charles V at Ingolstadt on 3 September 1546, who subsequently served under Philip II of Spain and returned to Nürnberg in 1569, where he became Burgermeister and Councillor; in his youth he had been a pupil of the great writing-master Johann Neudorffer, and later in life wrote an elegant calligraphic autobiography, illustrated by himself (Coburg, Landesbibliothek, MS 11).[10]

Melchior had an even more famous younger brother, also called Sebastian Schedel (1520–47), whose promising career as a doctor and botanist was cut short when he was murdered, by Spanish soldiers, it was thought, in the woods between Erlangen and Nürnberg as he was returning to the city.[11] It may have been from him that Sebastian inherited his interests, but he never knew his uncle, for his father only married late in life, in 1563, presumably when he retired from military service. His wife was Margaretha Reich, and over the next six years they had four children. Sebastian was the last, born on 9 October 1570.

There is a manuscript account of the Schedel family in the Nürnberg Stadtarchiv, which provides a long and revealing account of our Sebastian.[12] Of his early life we know nothing, but in 1587 his guardians sent him to Prague, where he became secretary to the 'Reichs Hofraths Secretarius', Dr Ludwig Haberstock. After five years of satisfactory service he moved to Vienna in 1592, to work for Dr Christoph Prickheimer zum Hundsthurn, Imperial Councillor and Secretary for Upper Austria, whom he served until 1597. In 1599 he embarked on a long journey at his own expense to Hungary, Slavonia, Dalmatia, Croatia, 'Siebenbürgen' (Transylvania), the Moldau, and adjoining countries, including 'ganz Italien'. He was able to travel at his own cost, 'supporting himself by the art of painting, in which he was very skilful'.

His Italian journey was in 1606, and it included a visit to the Venetian territories with his friend Seifried Pfinzing von Henfenfeld. In 1608, like his father, he returned to Nürnberg, taking service with the city as an ensign, under the captaincy of Wilhelm Groland. Next year he married Maria Salome Rieter, by whom he had two children; she died in 1613, and he married again in 1615 to Susanna Lucia Alstadt, and had two more children.

In 1608 also he became 'Ochsenamtmann', or Steward of the Butchers' Guild, and held down the office until 1616, but as he employed all the Guild's resources, and spent too much money and time, 'auf Mahlerey', on painting, to its no little loss, he was removed from his office. He spent the rest of his life painting, and supported himself by the produce of his brush, 'seinen Besengüter'. This did not prevent him from taking part in the proceedings on the solemn entry of the Emperor Matthias into Nürnberg on 2 July 1612, where he stood on the wall in front of the Spittlertor with the 9th company of the citizens, as ensign under Captain Lazarus Heller. He continued to take part in the city's affairs, and his name appears in the 'Libri Litterarum', or summaries of evidence in law cases relating to property, as testifying on eight occasions between 1609 and 1624.[13] In later life he suffered much from the stone. He died on 13 March 1628 and is buried in the Johanneskirche.[14]

The book at Kew is not the only example of what we must now consider as Sebastian Schedel's own work, with the exception of the cut-out engravings and, perhaps, the drawings similarly treated. An even more sumptuous example of his work is the 'Schoenbartbuch', an illustrated chronicle of the participation of the Nürnberg Butchers' Guild in the Fassnacht celebrations.[15] It begins with both verse and prose texts of a chronicle of the origins of the

---

10. Ilona Hubay, *Die Handschriften des Landesbibliothek Coburg* (Coburg, 1962), 29–31 (MS. 11).
11. G. A. Will, *Nürnbergische Gelehrte-Lexicon*, III. 502.
12. Stadtarchiv Nürnberg, Handelvorstand E8 Nr 5138, 'Das Schedelsche Geschlecht', ff. 11–12.
13. Stadtarchiv Nürnberg B. 14/I, 'Libri Litterarum', CXII, 215, CXIX, 23, CXXI, 14, 81, CXXII, 40, 237, CXXIV, 161, and CXXXIV, 94.
14. Carl L. Sachs, 'Metzgewerbe und Fleischversorgung der Reichstadt Nürnberg', *Mitteilungen des Vereins für die Geschichte der Stadt Nürnberg*, XXIV (1922), 24, 125.
15. University Reference Library, University of California at Los Angeles, Special Collections, MS. 170/351. The manuscript, one of a group from the Liechtenstein collection, was acquired by the Library from H. P. Kraus, New York, through the Samuel H. Kress Foundation. It is listed by Hans-Ulrich Roller, *Der Nürnberger Schembartlauf: Studium zum Fest- und Maskenwesen des spätten Mittelalters* (Tübingen, 1965), 236.

Schempartlauf. Then follows the main part, a series of entries covering the years 1449–1539, with an addition for 1600 (due to its violence, the Schempartlauf was suspended in the interim). Each entry lists the participants, from all the noble and patrician families of Nürnberg, the 'Laufer' or masked guard for the Butchers' Guild during their dance,[16] and the 'Hoelle', the elaborate float or sled with a different tableau scene for each year. Finally, there are the 'Grotesques', figures of pure fantasy that attended the Schempartlauf. The book ends with five double page panoramas of various Fassnacht celebrations.

The vigour and skill of the many paintings show Schedel's talents in a new guise. It is a sumptuous book, bound in black morocco gilt with green ties, which well explains the expenditure which the Butchers' Guild ultimately found too much. Like the book at Kew, it has Schedel's painted bookplate, undated, but facing it is a handsome portrait engraving of Schedel, plainer but similar in style to that of Bishop Johann Conrad, this time by Lukas Kilian, dated 1614.

What, then, was the relation of this figure to the *Hortus Eystettensis*? He must, I think, have been involved in the enterprise from the outset, or a year later, when he returned to Nürnberg. His Italian journeys (and where are the pictures he made then?) may have commended him to the Bishop, who had travelled there extensively in his youth. Did he go to Eichstätt and make drawings there, or did he paint from the specimens sent to Nürnberg in boxes? The set of detailed studies of individual flowers of 'Iasminum indicum'[17] suggests the latter, but other drawings must have been made from live plants. Are any of these the actual drawings from which the 'plate' drawings were composed? The answer is clearly no, since none of them have the curious recognition mark to correspond with the marks on the 'plate' drawings. At the same time, however, Schedel must have been closely associated with the colourists, since he was able to pick up fragments of several sheets of the *Hortus Eystettensis* (including two copies of one), discarded for some reason during the process.

The conclusion must be that Schedel's drawings, which were clearly his own property, are for the most part an early stage of the process by which the Bishop's flowers found their way into print. They are not perhaps the earliest, although the sheets now bound in Schedel's book may once have been loose and taken in an artist's portfolio to Eichstätt; and they are certainly not the latest — these, unlike the 'plate' drawings, did not survive the final dissolution of the last colouring workshop (perhaps the 'plate' drawings survived only because they were already bound, although the present binding was provided by Trew). What is particularly interesting is the more stylized, less naturalistic, drawings, where Schedel appears to be adapting his style to a pre-existing formula, one perhaps originally devised twenty years earlier by Camerarius.

Schedel cannot have been the only artist involved; who the others were is one of many unanswered questions. We know from Hainhofer's letters to Philipp II of Pomerania that Daniel Herzog, his own favourite flower-painter, was conscripted to the task.[18] The named artists who painted the church roofs described earlier came from Nürnberg, and, as specialists in flower painting, may also have been among those involved. One of Schedel's pupils was later, and perhaps also earlier, involved. But names are few and far between. We do not know who actually converted Abbot Johann V's careful drawings into the painted decoration of the Michelsberg vaulting. We can only guess at the process by which other drawings from nature were converted into the plates of the *Hortus Eystettensis*. It may well be that further examples of the whole corpus of drawings generated by this enterprise may be found elsewhere, in collections of drawings, artistic or botanical, unrecognised as such. A number of extremely fine drawings of flowers, mainly tulips, on vellum, is currently in the stock of Ursus Books, New York. Clearly German, on the evidence of the captions in contemporary manuscript, but perhaps executed later in the seventeenth century, to judge by the style, they may derive directly from drawings made for the *Hortus Eystettensis*, rather than from a coloured copy of the book itself. Perhaps the most interesting link in the chain, the most important of all, is how the drawings for the plates were actually laid out. The 'plate' drawings are an amazingly skilful adaptation, converting drawings of as

---

16. Samuel L. Sumberg, *The Nuremberg Schembart Carnival* (Columbia University Germanic Studies, vol. 12, New York, 1941), 57.
17. 'Kalendarium', f.176.
18. Doering, 235 (see Appendix A, p.63). Herzog may, in fact, have been engaged on the work of colouring printed copies, rather (at this late stage) than original drawings. See below, p.51 (Leufsta).

many as eight separate flowers (each one perhaps represented by more than one 'original') into a single composition. But someone not the 'plate' artist must have made the selection. Although the plants appear in seasonal order and similar plants — notably the various tulips and irises — are grouped together, grouping is not universal. Often pages contain flowers quite different in size and habit. The same plant may occur twice, in different contexts, sometimes widely separated. Was this arrangement purely aesthetic, or did it reflect, to some degree, the pattern and order in which the Willibaldsburg was planted? Or, again, did it reflect some higher, philosophic, order, something once discussed by Camerarius and the Bishop? These are questions that cannot now be answered, but we are left with the certainty that the impresario's role, of which this would have been part, was no light task. Besler, if it was he who undertook it, deserves greater credit for the final achievement than the academic botanists of the later seventeenth and eighteenth century were prepared to allow him.

COLOURING THE BOOK

The early coloured copies

LANNING the *Hortus Eystettensis*, as Bishop Johann Conrad had entrusted it to Besler, had, like the garden itself, been an enterprise on a grand scale. It had involved two editions, one consisting of the plates alone on superior paper designed for colouring, the other on plainer paper with a letterpress text. The first was the true realization of the scheme that went back to the 'Florilegium' prepared by Camerarius, a pictorial record emphasizing in colour the distinctive features of all the different plants, a visual summary of this whole branch of creation. The other was destined for a more professional market, botanists, naturalists and the learned world generally, for whom the printed references would be an added and useful guide to other printed works with which they would be familiar. These copies must always have been intended to be circulated through the established commercial network of the book trade, in particular the Book Fair at Frankfurt, the great international entrepôt, and it was perhaps originally only through these that Besler intended to make his 'Ruhm und profitt', his fame and fortune.

The coloured copies, then, may have been intended to be exclusively for presentation by the Bishop, to provide an answer to the requests for 'Contrafetten' that others besides Herzog Wilhelm V might make, as well as to strengthen links, horticultural as well as diplomatic, with neighbouring potentates like Joachim Ernst, Markgraf of Ansbach. It may, on the other hand, always have been part of the project that they should have been available for sale. The price at which they were sold, 500 florins, must, in the event, be an accurate reflection of what they cost to produce, over and above the 'base cost' of paper and printing, reflected by the selling price of the plain edition, which, as we have seen, varied between 35 and 50 fl. The scale of this enterprise, that is, the amount of skilled time (and the not inconsiderable amount of materials), must have required, at the outset, elaborate negotiations with those capable of doing the work.

It is highly probable, therefore, that Besler, in addition to arranging for supplies of the two papers, for the engraving and printing of the plates, and the letterpress printing, made arrangements with a professional 'firm' of illuminators. One such existed in Nürnberg, where it was active for a century or more, that of the Mack or Mackh family.[1] Hans Mack is recorded as active between 1536 and 1582 as 'Briefmaler, Formschneider und Illuminist'; his name also appears on two Nürnberg imprints in 1579/80. Georg Mack 'der Älterer', so called to distinguish him from his son, may have been Hans's son. His work is recorded from 1556; he painted several leaves in the 'Stammbuch' of Hieronymus Cöler (B.L. Egerton MS 1184), and a collection of woodcuts by Virgil Solis (Coburg, Kupferstichkabinett). He also worked with Jost Amman on the Pfinzing Bible (recorded as on loan to the Germanische Museum, Nürnberg) and on a prayerbook in the collection of Otto Schäfer, both dated 1568. He died in 1601.

It was his son, Georg Mack, originally 'der Jungerer', who was entrusted with the colouring of the earliest recorded copies of the *Hortus Eystettensis*. In 1582 he obtained permission to set up his own workshop. His earliest recorded work is in an account book, whose first two pages are illuminated and signed, respectively, 'G.M.I.' and 'Georg Mack Iung: Illuminiert 1585'. Shortly after, he executed the decoration of a fine set of the Plantin Polyglott Bible, formerly in the Doheny collection,[2] most of it signed in unusual detail. He describes himself again as 'Iunger' or 'Iunior', and 'Hats:', perhaps an abbreviated place-name, of birth or residence; the first volume is dated 17 and 18 April 1587 by the new kalendar ('neuen Calē.'), and the last 1588. By the time that he became involved in the *Hortus Eystettensis* his father was dead, and there was, it seems, a grandson, also called Georg, since his father's

---

1. The details given below come mainly from the entries in Thieme-Becker, *Allgemeine Lexicon der bildenden Künstler*, XXIII, 518, and J. Benzing, *Buchdruckerlexicon des 16. Jahrhunderts*, 1952, 140 (no. 76).
2. Sold at Christie's, New York, 17–18 October 1988, lot 1066.

signature alters to 'G.M.A' or 'G.M. Seni.' It would be agreeable to associate the single plate signed 'G.M.I.' with the third Georg, but the 'I' may only mean 'Illuminist' or 'Illuminiert', as elsewhere.

But the Mack family were not the only artists employed by Besler in the enterprise. Georg Schneider was clearly an artist of equal ability, and perhaps the senior, or thought to be the more able, since to him fell the distinction of colouring the title-pages of the first two copies to be completed (so far as we know), those at Nürnberg (N) and (apparently) divided between Turin (T) and Leufsta (L).[3] This sumptuous page is made more so by the deliberate attempt to imitate shot silk on the title-cloth in N, and the elaborate reverse in T; the dark blue background suggests a primal night out of which the gate leads to the Garden of Eden. The care with which the plates have been coloured is both elaborate and remarkably even. The quality, technique and colours used are extremely similar in N, T and L. The technique, despite the large size of the leaf, remains that of the miniaturist, with fine hatching used both as a tint and a means of shading.

The other artists are only identified in their work by monograms and initials. 'MB' only appears in N, and 'DR' in N and T. 'HS' in both N and T links suggestively with 'HSF' in the copies at Munich (BSB) and Vienna (OSB), the latter of which is firmly dated late in the century. 'GH' and 'DH', which are found in L, may stand for Georg Gärtner (Hortulanus) and the elusive Daniel Herzog. 'IN' and 'RTAF' appear only in the copy at Madrid (BNM). In the Vatican copy (V) the engraver's signature 'RC' (Raphael Custos) is twice painted in. Finally, both BSB and OSB contain the initials 'MF' and also the signature of Magdalena Fürstin, of whom more later. Only the King's Library copy in London (BL) has almost all the plates signed, and several dated. The frequency of signatures in the other copies is surprisingly varied; it is best studied in the table (Appendix C, p.73).[4]

Why, it may be asked, are some but not all the plates signed by the artists? It is very difficult to answer this question: if the Water Lily (111i) or the 'Ficus Indica' (359) might seem spectacular and their colouring worth special signature, why is the great double plate of the 'Martagon Imperiale' (181–2) not similarly distinguished? Perhaps there are other reasons, such as the need to chart the progress of individual stints of work in a co-operative enterprise. Pride in achievement may well have been there, but when remuneration was calculated on a piece-work basis a signature had a more practical utility in indicating the hand into which payment should fall. It would be easier to make some guess as to how the work was assigned (and whether signatures indicated the beginning or end of a given stint, perhaps), if we knew the order in which it was done. Such evidence as exists, the system of numbers on the 'plate' drawings (ErZ), for example, suggests that it did not follow the published sequence; more cannot be surmised.

The first two copies to consider are N and T + L. The dates that they bear indicate that they come first in the sequence of coloured copies, so far as it can be established; the reasons for linking T and L are set out below (pp.50–51). Both copies have the printed text on separate sheets. The two sets of plates were clearly painted in parallel, the same artists for the most part (although not invariably) undertaking the same pair of prints. It is impossible to say whether the plates were painted in sequence; the full date, 4 May 1613, on the 'Ficus Indica' (plate 359) in both probably (in this instance) only reflects the size, novelty and importance of the great plant at Eichstätt.

The date '1614' on the season-title(s) is interesting since it shows that the book for Nürnberg could not have been ready before that date. According to the 'Extractus' it was not 'purchased' (or perhaps paid for) until 1617; this may be a simple error, but, as we have seen, that year saw the conclusion of the enterprise in a number of ways, perhaps including a final settlement of accounts with the Nürnberg Council. On the other hand, Besler rendered an account of some sort to the Chapter at Eichstätt on 23 September 1613 for their copy. What the force of the word 'einzige' is in this context is not clear from the 'Extractus'; it can hardly mean the only copy of any sort, since the Chapter requested 25 plain copies a fortnight later; perhaps it means, as we would say, 'unique'.[5]

The interesting point about both these transactions is the enormous price, 500 fl., that both bodies evidently expected to pay for their copies. The Herzog August at Braunschweig may have regarded this as an exorbitant sum,

---

3. Georg Schneider, 'Formschneider und Briefmaler' (wood engraver and illuminator), is a recurrent figure in the records of the Nürnberg Rat between 1583 and 1621 (Hampe, *Nürnberger Ratsverlässe*, II, *passim*).
4. 'M.B.' may be the 'Goldschmied und Kupferstecher' (goldsmith and engraver on copper) Mathias Beytler, *fl.* 1582–1614 at Ansbach (Thieme-Becker, *Allgemeines Lexicon*, III, 556).
5. See above, pp.15–17.

but the Nürnberg Council and Eichstätt Chapter, both of which might have considered Besler as considerably indebted to them, seem to have accepted the price as fair, and one which they were due to pay. The binding, in both instances (that is, the Nürnberg and Leufsta copies, which closely resemble each other), seems have counted for less than might be expected. Although distinguished, they were not by current standards especially sumptuous.

The fate of these two copies is discussed in detail below (see pp. 50–51). It seems highly probable that the copy now at Nürnberg has never left the city. It may have been transferred to the Stadtbibliothek by the Council at some point; certainly, the copy coloured by Magdalena Fürstin (1652–1717) is first recorded there by Doppelmayr in 1730 and was still there when Hirsching published his account of the most remarkable German libraries in 1788, in which it is described as bound in two volumes; it is now lost.[6] Presumably the Mack copy was still there when the Council acquired the collection of Adam Rudolph Solger (1693–1770), in 1766, since it was thus available to fill a clearly painful gap on Solger's shelves.

The Eichstätt copy seems to have had a more adventurous life. It had clearly been lost to Eichstätt long before Starckmann and Trew began their investigations of the past history of the great book. It seems all too probable that it formed part of the booty of the Swedish troops under Bernhard von Weimar.[7] How the two volumes came to be divided can only be surmised, but the first volume clearly stayed in Germany, perhaps damaged and therefore in need of rebinding later in the century. In the eighteenth century it was in the hands of David Samuel von Madai, and presumably remained in Halle after his death in 1780. There is no reason to suppose that it left Germany until recently. The survival of the other volume and its acquisition by Olof Rudbeck is perhaps an even greater miracle, but since its arrival in Sweden its history is clearly documented, from its acquisition by Charles De Geer from the younger Olof Rudbeck to its final transfer to Uppsala University.

The next copy whose colouring can be securely dated is that now in the British Library, with no printed text. The chronological progress of the colouring of this copy can be followed to some extent, from 2 March 1614, the date of plate 78, to the titlepage, 16 April 1615. The bulk of the dated plates follow plate 78, all dated '1614'. All are signed by Georg Mack, with various flourishes attached to the name, 'A' for 'Älter' or 'L' for 'Luminist', and also with some variant spellings, such as 'Ioerg'; none of these suggests any division of work, except possibly with his son. This may indeed indicate that BL is his entire and sole work. It may, equally, be that the co-operative, each member settling separately with the paymaster, had ceased to exist, and that whether or not he executed the work Mack signed for it as master of the workshop to whom payment was due. Again, there is no correlation between the difficulty or elaboration of the work and the incidence of signatures, apart from the full signature on the title-page; the 'Ficus Indica' plate is not signed. The German text is written in large calligraphic script on the verso of each plate, indicating the plants on the facing plate.

The quality of the colouring, the frequent use of gold, the technique of fine hatching in a contrasting colour over a base, make it clear that this was, by any standards, a *chef d'oeuvre*, and the extraordinary range of signatures suggests that it was intended to be so considered. But who the patron was for whom it was executed cannot be determined. The most obvious would be Wilhelm V of Bavaria (1548–1626), of whose interest we have Hainhofer's evidence, or his son Maximilian I (1573–1651), who was even more interested in horticulture as such.[8]

The next two dated copies from the Mack workshop are the first copy in the Staatsbibliothek, Berlin, and that in the Biblioteca Apostolica Vaticana (B1, without text, and V, with text). Both are more sparsely signed, and each dated in one place only (plates 169 and 78). The Berlin copy is simply dated 1615; the Vatican copy 4 June 1615, if my reading of a lead-point scribble, an unfinished signature, is correct. In both copies the colour scheme corresponds closely with the other earlier Mack workshop products, although with a slightly diminished palette and a tendency

---

6. Johann Gabriel Doppelmayr, *Historische Nachricht von den Nürnbergischen Mathematicis und Künstlern* (Nürnberg, 1730), 270–1; Friedrich Karl Gottlob Hirsching, *Versuch einer Beschreibung sehenswürdiger Bibliotheken Teutschlands* (Erlangen, 1788), III.i, 67.
7. Swedish interest in cultural loot was not to be underestimated. Herzog August wrote to Hainhofer in 1636, thankful that he had moved his library to Braunschweig: 'dass ich meine Bibliothec, vor etwa 14 tagen, anhero transferiret: dan die Septentionales ein böses Auge darauf gerichtet: Ich habe sie selber anhero beglaitet'. See Jill Bepler, *Ferdinand Albrecht, Duke of Braunschweig-Lunenburg (1636–87): a traveller and his travelogue* (Wiesbaden, 1988), 48.
8. B. Volk-Knüttel, 'Maximilian I. von Bayern als Sammler und Auftraggeber: Seine Korrespondenz mit Philipp Hainhofer 1611–1615', *Quellen und Studien zur Kunstpolitik der Wittelsbacher von 16. bis zum 18. Jahrhundert* (München, 1980), 92.

towards conventionalization. The unusual signature in V also seems to be unfinished: the 'A' and 'H' in 'GAH' run together; most likely the signature was to have been Gärtner's, less so that it was yet another form of Mack's. 'M^9' is probably a conventional abbreviation for the latter. In neither case has the original binding survived, although the two volumes, by now bound in splendid Roman red morocco, of V may represent an earlier division, and B1 may have always been a single volume. The printed text in V was obviously present *ab initio*; the manuscript text in B1 is equally clearly an afterthought, though not long after.

There is no way now of telling for whom these copies were originally made. Bishop Johann Conrad had unusually close contacts with the Markgraf Joachim Ernst von Brandenburg-Ansbach, but there are other routes by which a copy could have been acquired then or later by the future Prussian Royal Library. The Kurfürst Friedrich von der Pfalz (1583–1610) visited Bishop Johann Conrad in 1599, and it is possible that V was originally destined for his son Frederick V, the hapless Elector Palatine and husband of the 'Winter Queen'. The fact that it was commissioned from the Mack workshop makes it virtually certain that it was done to a German commission; on the other hand, there is no sign that it was part of the 'spoils of war' by which the Bibliotheca Palatina reached Rome in 1622.[9] Such a book might well have engaged the interest of Pope Urban VIII, or have been given by him to his nephew Cardinal Francesco Barberini, whose 'museo cartaceo', created by Cassiano dal Pozzo, was far greater than any of the collections of 'Contrafetten' of similar natural curiosities formed by German or Austrian princes. It was, at least, certainly in the Barberini collection by the second half of the seventeenth century, and passed, with the rest of the great Biblioteca Barberini, to the Vatican Library in 1902.

Another copy that clearly stems, at least in part, from the Mack workshop is that in the Biblioteca Nacional, Madrid. Like the Vatican copy it has come a long way from an original German location, but in this case there is no trace of what that might have been. There is no date at any point. The present binding in two volumes of unequal size may not represent an original arrangement; the copy has clearly suffered some damage, either before or after reaching the Biblioteca Real. The pattern of colouring is not in itself remarkable, though it provides an interesting link with the copies of early German provenance but without any links with the Mack workshop, notably the copy now at Eichstätt. The signatures of the plates have only the slightest correspondence with the other copies, one each with B1 and V; it has a unique signature on plate 336, which in other copies is signed by Mack or Schneider.

The last copy that can be connected with the Mack workshop is that in the Hessische Landesbibliothek at Wiesbaden (W). In this, only the title-page has been signed and dated 1617. This is more than a year later than the previous copies (unless their colouring continued into 1616; the title-pages, which might have provided a *terminus ante quem* for completion, are undated). The year 1617 also saw some conclusion to the initial stage of the enterprise, which may have involved the Mack workshop. W has the portrait of Besler, but not his arms, and no text. None of the plates are signed, and the colouring, though its scheme remains close to that of other products of the workshop, is by comparison limited in palette and less detailed in execution. Although the copy is still in its original binding, it has been substantially repaired, and any sign of its early provenance has disappeared.

It remains, now, to consider the four other copies that constitute the 'early' group, which were not coloured by the Mack workshop. Only one of these bears a date, the second copy in the Staatsbibliothek, Berlin (B2). This is the best documented copy of all in terms of provenance, since it was presented to Philipp II of Stettin and Pomerania in 1617 by Philipp Hainhofer, his hopes unfulfilled that the Bishop would give his patron a copy or that his patron would buy one. The copy is remarkable in that it is one of the 'plain' issue, with the printed text on the verso. The lower quality evidently gave the colourist or colourists problems. The paint had to be applied sparingly and quickly to avoid it spreading or permeating to the other side of the sheet. In some cases, inevitably, the colour did spread, and the outline was 'corrected' with a painted dark red line; the bulbs are similarly treated. The title-page resembles the later Mack copies with a pale background and thick black border; the Bishop's arms are correctly reproduced, but, on the portrait page, Besler's are not.

---

9. J. Bignami Odier, *La bibliothèque Vaticane de Sixte IV à Pie XI* (Studi e Testi 272, Città del Vaticano, 1973), 107–10.

It looks as though Hainhofer, finally driven by the prospect of losing a copy of any description for his favourite patron, but unwilling to pay Besler's price for one properly coloured on superior paper, had bought one of the plain copies, for which he would have had to pay something like the 48 fl. he quoted to Herzog August of Braunschweig, and had it coloured without reference to Besler. His colourist or colourists must have had access either (privily) to the drawings or colouring guides used by the Mack workshop, or to a coloured copy already made. Perhaps the latter is the more likely, although it seems unlikely that any of the extant copies was used. There are two places where red is introduced incorrectly, notably for the flower of 'Papaver spinosus' (Plate 288i). This error recurs in some of the later copies, which may be due to continued access to an independent source or simply to a 'conventionalizing' tendency (berries and poppies ought to be red).

The copy formerly in the De Belder collection (DeB) is different in every way from B2, except in its lack of any visible association with Besler. It bears all the signs of one of the superior copies — 'watermarked' paper, no printed text, the German common names inserted in manuscript, but the portrait and arms of Besler are conspicuously absent, and also his address 'Ad lectorem'. The colouring of the title-page corresponds generally to the Mack pattern, but the background is grey-brown, strongly suggesting a wall in which the gateway to a paradise is set, and this in turn is surrounded not by a black border but by paper coloured black as far as the trimmed edges; the title-cloth is coloured yellow, with only a suggestion of red shading, painted, not hatched. There is no gold.

The plates are painted in much the same scheme as the Mack copies, but with a much more modern sense of colour values. Instead of deceiving the eye, building up the appearance of colour by the use of two quite different shades in contrast, a whole range of 'real' colours are used, the light and dark shading achieved by varying the colour, with almost no use of high-lighting. In this respect, the colouring bears a resemblance to the more finished of Schedel's drawings. From a botanical point of view, the colour is almost always more accurate than the 'Mack version' when the two diverge. In the list of variants (App. B, p.68), DeB shows a greater number of unique 'readings' than any other copy, and most though not all of them are right.

The other unique feature of DeB is in the German common names. Not merely the spelling but the text of these differs from those of BL, B1 and Er, which in general correspond with those in the printed text. Not surprisingly, and for the same reason that the colours are sometimes transposed, the captions in these others are occasionally applied to the wrong plants. In DeB the captions are correctly placed. The binding is unusual. Although the use of tawed leather continued until into the eighteenth century, the use of overall tooling would have been somewhat old-fashioned by the second decade of the seventeenth century. The pattern of tooling, as well as the choice of leather, was unusual.

All that can be derived from this is that the copy was evidently painted for a prince with a strong and well-based interest in horticulture (or access to such an interest), clearly familiar with Bishop Johann Conrad's garden as well as the book based upon it, but with independent access to the sources of reference for colouring the latter, unconnected with Besler or Mack. The colourist did not need to sign his name, his identity evidently clear without saying. The further clue to the identity of the patron comes from the inserted drawing and the print based upon it, relating to the growth of an Agave plant, first at Nürnberg, and then at Ansbach, 26 miles (42 km) away, where it finally flowered in 1627. It was the first to flower in Germany, although its form was not unknown since another plant had flowered in 1586 in the botanical garden of Francesco de' Medici, Grand Duke of Tuscany, who had had it painted by his court artist, Jacopo Ligozzi. A copy of this had been sent by Giuseppe Casabona, curator of the garden, to his friend Joachim Camerarius the younger, who published it in 1588 in the *Hortus medicus et philosophicus*. Still, the growth of this immensely tall shoot, whose flowering was of legendary rarity, caused a sensation, as recorded in the caption to the print. No doubt drawing and print (itself now a great rarity) were thought to be appropriately preserved in the *Hortus Eystettensis*, as a compendium of up-to-date botanical illustration.[10]

---

10. Besides this copy, there is another example of the print in the Herzog Anton-Ulrich Museum, Braunschweig, the copy supplied originally to Herzog August (see below, p.66).

The most likely person, therefore, to have commissioned this copy was the builder of the handsome *Lusthaus* with its attached gardens at Ansbach, Joachim Ernst, Markgraf of Brandenburg-Ansbach (1583–1625), ruler of the principality from 1603. The details of the garden can be clearly seen in Wenceslaus Hollar's print of Ansbach, made about the middle of the seventeenth century (reproduced opposite). The Agave itself probably grew in the 'Pomerantzen Garten', lettered 'R' in the key, but not so indicated on the view itself; the orangery may have been the building to the right and at an angle to the *Lusthaus*. The long and vivid description of it in the print makes it clear that the plant was brought from Nürnberg to Ansbach in 1610, when it was already twelve years old, and placed in the prince's *Lustgarten*, and was carefully tended there until, on 4 July 1627, its prodigious growth began, ending the following spring in a head of 180 (or more) flowers.[11]

The Markgraf did not live to see this splendour, for he died on 7 March 1625. But no doubt it was his interest that suggested the commemoration of this notable event with the drawing and, subsequently, Wolfgang Kilian's print. It may be that the artist of the drawing was also the colourist of the copy of *Hortus Eystettensis*. If so, he must have had access to the Nürnberg colourists' material, probably through Besler, who would have been familiar with his old patron's close relations with the Markgraf, and may indeed have supplied the Agave plant, whose growth is dated from the year that Camerarius died.

Joachim Ernst's life was mostly spent in the field, where his generalship was admired, but his library catalogue shows him to have been a man of considerable learning and wide interests, both in religious works and on such subjects as hunting, botany, medicine, astronomy, architecture, military tactics and fortification, and tournaments (the tilt-yard can be seen in the print to the right of the palace). He read French, and his library contains a number of French historical works, as well as the works of Duplessis-Mornay, which reveal his Calvinistic religious sympathies. It also contained a copy of the *Hortus Eystettensis*, listed as '2 Tomi – Hortus Eystettensis'.[12]

It is possible that this copy remained at Ansbach until the abdication of the last Markgraf Alexander in 1791, in favour of his cousin, Friedrich Wilhelm II, King of Prussia. After some hesitation, during which part of the collection was transferred to Berlin, the major part of the library was transferred to the University of Erlangen, founded in 1743 by Alexander's father, Friedrich, Markgraf of Brandenburg-Bayreuth. Three copies of the *Hortus Eystettensis* came to Erlangen in October or November 1805. One of these, an uncoloured copy of the '1713' edition, is still at Erlangen; the other two are marked in the catalogue 'given away on 1.7.1839'.[13] By this time, in the late autumn of 1818, to be precise, the library of the University of Altdorf had been merged with that of Erlangen. With it had come the famous collection of Christoph Jakob Trew and, with that, the even more famous coloured copy of the *Hortus Eystettensis* that had been at Altdorf since 1696. This superfluity may have suggested the disposal of the two 1613 copies from Ansbach, one of which may have been this, less regarded because less well-known. It is even possible that Bernard Quaritch (1819–99), not yet twenty, may have handled it, as an apprentice bookseller with C. H. Beck at Nordlingen. Presumably it remained in the trade in Germany[14] until sold by Baer at Frankfurt to Maria Theresa Earle.

The copy now at the University Library at Erlangen demands consideration next. This is remarkable as the only coloured copy with an unbroken pedigree going back almost to the beginning. The title-page, with its slate-coloured background (without black border, however) and yellow title-cloth (the green reverse is like T and V)

---

11. The memory of the event was no doubt dimmed by war, and a learned discussion of the priority of Agave flowerings in the *Miscellanea curiosa sive ephemeridum medico-physicarum Germanicarum academiae naturae curiosorum decuriae 1* for the year 1670 (Frankfurt/Leipzig, 2nd ed., 1674) gave it to Stuttgart (1658–9); this was put right in the 'Appendix ad annum decimum decuriae II' in 1692, where the 'Aloe Onoldina' of 1627 is correctly described, and its second flowering in 1687 (pp.56–8). The facts were definitively set out by Christoph Jakob Trew in *Beschreibung der grossen Americanischen Aloe* . . . (Nürnberg, 1727). For an excellent modern account, see Brigitte Volk-Knüttel, 'Eine blühende Agave im Münchener Hofgarten 1634', *Aufsätze zur Kunstgeschichte: Festschrift für Hermann Bauer zum 60. Geburtstag*, ed. K. Mösender & A. Prater (Hildesheim, 1991), 194–205. See also below, Appendix A, p.67.

12. The inventory *post mortem* of Joachim Ernst's library (Staatsarchiv Nürnberg Geheimregistratur, Bamberger Zugang Nr.51) lists it thus. See G. Schuhmann, *Ansbacher Bibliotheken vom Mittelalter bis 1806* (Schriften des Instituts für Fränkische Landesforschung an der Universität Erlangen, Bd 8) (Kallmünz, 1961), 85f. One of the Markgraf's characteristic bindings is illustrated (plate 1). I am grateful to Archivoberrat Dr G. Richter for verifying the exact reference in the 'Nachlassinventar'.

13. K. Wickert, 'Die Erlanger Exemplare des "Hortus Eystettensis"', *H.E.*, 121.

14. *Ibid.*, 138–9.

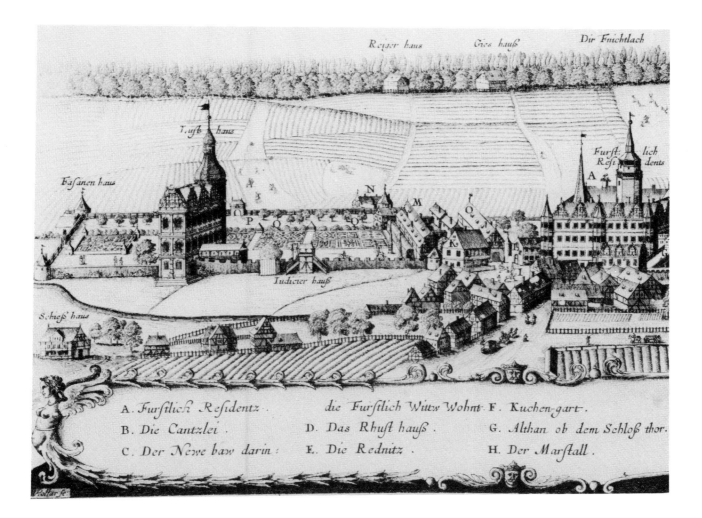

associates it with the primary stage of colouring, although it has neither the precision and variety of that in DeB, or the sumptuous elegance of the copies signed by the Mack workshop. The same is true of the plates, which reproduce the consensus of the early copies with remarkable fidelity and uniformity. Like BL, DeB, B1 and W, it has no text, and like them (though not uniformly) it has the German common names in manuscript. It is, in short, another sober if less imposing copy, such as one of Besler's colleagues in the Collegium Medicum, Caspar Hofmann, the Dean, who contributed the verses to Gärtner's 'Zierblatt' for the Markgraf of Ansbach, might have commissioned. It may be significant that, like W, it has the portrait print of Besler, but not his arms.

It is not surprising, therefore, to find it in the possession of Johann Leonhard Stöberlein, born in Nürnberg in 1636, a student of medicine and botany at Padua, like so many of his contemporaries, and from 1660 to his death in 1696 in practice as an apothecary in Nürnberg.[15] He was also a poet, and in that capacity was admitted with his wife to membership of the 'Pegnesischer Blumenorden'. On his death he bequeathed his professional library to the University of Altdorf, where it remained a valued and much advertised treasure,[16] until the University of Altdorf was closed in 1809 and its library transferred to the newer University of Erlangen in 1818. There it has been ever since.

The last of the 'early' copies is that at Eichstätt. It too has no signatures of colourists or other distinguishing marks. The background to the titlepage is, unusually, dark brown, with the familiar black border. The title cloth is yellow (red reverse), and the names on the pedestal are left as black on white, with the edges shaded, instead of being coloured blue. The colouring of the plates is hasty and sometimes imprecise, with a relatively restricted palette; it has no distinctive or unique characteristics, and the details of the individual plants conform to the other 'early copies'. The only unusual feature of the book is the double plate of the portrait of Besler and his arms, which have been reversed. No other coloured copy has this feature, although it may appear in plain copies. The printed text is present.

---

15. *Ibid.*, 123.
16. It was kept on a table in the Library, with a special desk on which to read it, clearly visible in a mid-18th century print of the Library (*H.E.*, 125).

There is every sign that, like Er, this was coloured for a professional owner of some substance, and one who, unlike the first owner of Er, felt some need of the text. The present binding, although German, is clearly later, and one cannot tell if the copy was always thus in three volumes. Lord Aylesford may have purchased the book on the continent, *c* 1800. The rest of its pedigree is easily followed.

The 'early' copies, in sum, present a generally coherent group, divided in two by the presence or absence of 'Mack workshop' signatures. While two only of the 'Mack' copies can be associated, and that tenuously, with Besler, there is no reason to suppose that the others were not acquired direct from him, with the exception of B2. If, as seems likely, copies of any sort were then in short supply, Besler may well have been unwilling to release copies for colouring except at the full price. Like books printed on vellum such copies were generally supplied to order, and if Besler had a special arrangement for this with Mack that does not exclude the possibility that he would not have sold some whose colouring would be commissioned by the customer. The absence of signatures is, perhaps, less remarkable than their presence. There appears to be no special significance in the presence or absence of printed text, season-titles or the portrait of Besler, with or without his arms.

What, then, is common to the group? First and most important is the homogeneity of colour (with the interesting exception of DeB). Secondly, there is the general likeness in the treatment of the title-page, in particular the uniform yellow title cloth. The background, it may be noted, does change, moving from the dark and solid appearance of N, T and DeB to the paler background of BL, V and B2. Where one indicates a door from the temporal to the eternal, the other suggests a free-standing arch in a landscape, which you can as easily walk round as through, classical rather than baroque. Lastly, there is the volume of association, in terms of provenance, dates of colouring, and so on, going back to the first half of the seventeenth century.

There is one final characteristic that defines this group more precisely. It came to an end. There are no copies that conform so closely to this pattern datable after the mid-century, and everything suggests that the end came earlier, about 1630, with the death of Besler or the ravages brought to the area by the Thirty Years War. It is a striking, indeed unique, episode in the history of book production, and its success undoubtedly prompted the renewed and revived interest typified by the 'later' copies.

The later coloured copies

ALL the evidence suggests that the *Hortus Eystettensis* did not initially travel much outside Germany. The speed with which the plain edition sold out, the relatively local distribution of the early group of coloured copies — all this speaks for a limited circulation. Any further dispersal must have been severely inhibited by the Thirty Years War, which (ironically) resulted in two copies at least leaving Germany as war-booty. It is in the last year of the war, 1648, that the first coloured copy known to have left Germany by peaceful means, now in the Bibliothèque Nationale, reached its first owner. In the same year, Michael Rupert Besler produced his 'Mantissa' to the great work and dedicated it to the Archduke Leopold.

It is clear from the fragments of correspondence quoted earlier that, despite Hainhofer's gloomy prognostications, it was still possible to acquire a copy in 1628, when Michael Rupert Besler begged one from home. The family provided another copy for Altdorf University in 1630–1, and Francesco Pona's hopes in 1653 may have had some foundation. It is, however, interesting that one copy of the 1640 reprint produced by Bishop Marquard Schenk von Castell was prepared for colouring, even if the colouring did not get very far. This is the copy now in the Stadtsbibliothek at Mainz (M). While there is nothing to prove that the copy reached its present form before 1695 (the date of the manuscript index, itself only inserted), and its contemporary binding need not have inhibited later colouring, the balance of the colour variants suggests some degree of continuity with the 'early' tradition, even if others form a bridge with the later. In other respects, the copy is similarly devoid of indicative characteristics.

There was one other coloured copy of the 1640 edition in the Sächsische Landesbibliothek at Dresden. This may have been the same as one with a 'dedication' to Louis XIV, reported in the same library. Whether one or two, both are now probably destroyed.[1]

But the first datable 'post-war' or 'late' copy is, as noted, that in the Bibliothèque Nationale at Paris. This is inscribed on the title page by the owner, Charles Labbé de Montveron, and dated 1648. How and why a distinguished scholar, whose work on the Code of Justinian, in particular his editions of the *Novellae* and the *Basilica*, has given him a lasting place in the annals of legal history, came to acquire such a book is a mystery. His library deserves further study, the more so if, as seems likely, it was dispersed after his death.[2]

Although splendidly got up, with its elaborate rule borders and wealth of decoration extending to the initials and display lines of the text, this is not a copy designed for colouring but one of the 'plain' issue. Some of these may still have been available, if not from the Besler family then from the trade. Individual differences are not very marked, but the dark sky-blue background of the title now has no border, while the substitution of white for yellow for the title cloth suggests a significant break with the earlier or local tradition; the errors in the armorial bearings show that these were of no significance or importance now. The most significant change is the hardest to document, namely the restriction in the palette, suggestive of colouring from a key-chart, whether in visual (outline) or verbal form. But apart from rebinding this copy has suffered less than most from the passage of time, and its pedigree can be traced without a break from 1648 to the present day.

Much the same could be said of the other copy now at Paris, that in the Muséum de Histoire Naturelle.[3] This belonged to Guy-Crescent Fagon, Louis XIV's doctor, who began and ended his long career at the Jardin des Plantes. This copy differs markedly from all others in one important respect. While the colourist of the Bibliothèque Nationale copy met the problem of porous paper by sparing use of colour applied with a light brush, the colourist of MHN

---

1. K. Falkenstein, *Beschreibung der königlicher öffentlicher Bibliothek zu Dresden* (Dresden, 1839), 128, records a coloured copy of this edition. The 'editio illuminata des Hortus Eystettensis mit einer Widmung an König Ludwig XIV von Frankreich' is reported as once in the Sächsische Landesbibliothek by Theodor Neuhofer, 'Basilius Besler – Hortus Eystettensis', *Historische Blätter für Stadt und Landkreis Eichstätt*, VII.iii (1958), 10–11.

2. L. Michaud, *Biographie Universelle*, XXVIII, 338–41.

3. This copy has been reproduced in full, first in France as *L'herbier des quatres soisons* (Paris, 1987), then in Germany as *Der Garten von Eichstätt* (Munich, 1988), and finally in America as *The Besler Florilegium* (New York, 1988).

took the opposite course, using colours so dense that they could not spread, even though they might obscure the engraved detail. The titlepage is different from all the others; the sky-blue background (without border) is heightened with small white clouds, while the title-cloth is overpainted with its gold and silver lettering on purple. This too, like BNP, is a copy of the 'plain' edition, supplied by the same route. It too has an unbroken chain of provenance going back to the prime owner.

The copy sold at Monte Carlo (MC) is all the more interesting since it differs from both these copies in ways that suggest that it derives from an independent source, and is possibly anterior to them. In the first place, it was printed on watermarked paper, and was without text; such copies are less likely to have been available through the normal book trade than to have come, directly or indirectly, from the Besler family. The title-page resembles that of BNP is general appearance, but Cyrus's robes are reversed, the Bishop's arms are correct, and the broad border, if golden in colour, recalls the earlier style.[4]

Like the copy at the Bibermühle (S), MC looks both back and forward. It maintains some accurate 'early' detail neglected by BNP and MHN; it also contains some novel traits found in BNP, MHN and S. Its one completely unique feature, now sadly irretrievable, was the order of the plates and the notes, apparently intended to re-arrange the *Hortus Eystettensis* not according to the four seasons, but month by month, like Sebastian Schedel's 'Kalendarium'. This in turn must imply the existence of such a calendar; the form of the notes suggests that it recorded the flowering of some of the plants in a real garden in France about the middle of the seventeenth century. The style of the handwriting suggests a slightly earlier date, but, if the writer was old at the time of writing, it could be later, by which time sheets of the watermarked issue might have been more easily available again, as we shall see.

The copy in Switzerland (S) also fits Janus-like into the chain of transmission. The binding, a splendid piece of work in a style by now somewhat old-fashioned, was no doubt dictated by the first owner, Andrea Vendramin, whose arms appear not only gilt on the boards (elaborately made up with binder's tools) but also on a separate leaf before the title-page. The sheets are again on watermarked paper and without the text; they have hardly been trimmed, making this one of the largest copies known. The binding, although more elaborate and finished with gilt tooling, still resembles those of N and T, in structure as well as design. How such a book came to be bound in Germany for a distinguished Venetian collector, who might have been expected to bind it locally, and why — as the evidence suggests — it was not coloured there, are questions to which there is no obvious answer.

But despite all these links with the past, it is clear that, in terms of colouring, the sources of reference available to the artist of S were more remote than those of BNP, MHN and MC. The title-page arch has a golden hue, but the title cloth itself is white shaded red like BNP and MC, and the arms are wrong. The most interesting feature of the plates is the evidence of reversal (see App. E, p.79), evidence that the information on colouring was not in visual but in verbal form, and that in some cases at least the information, even when not misinterpreted, was wrong. There must have been two, if not three, stages of transmission separating S from the original sources available to Mack and the other early artists: the drawings must have been converted into uncoloured, perhaps 'key' outlines, with colour instructions; the colour instructions were reduced to a simple verbal list, keyed to the numbers of the plant on each plate in the *Hortus Eystettensis*; furthermore, either before, during or after this sequence, the original colour scheme given to some of the plants was altered. S remains a key witness to the chain of transmission. If any further copies belonging to this period or group should be discovered, their position in the chain can best be judged by reference to their similarity or otherwise to this copy.

Besler's last direct descendant, Paulus Basilius Besler, the last of the family to work at the famous Mohrenapotheke, died between 1651 and 1653. His death and the winding up of his estate may have released some copies of the *Hortus Eystettensis* not hitherto available. It is, however, perhaps more likely that it was the death of Besler's great-nephew, the son of Michael Rupert Besler, Joachim Hieronymus Besler (named after two famous forebears) that released a final residue. He died untimely, without heirs, in 1671, while a medical student at Altdorf. It is surely no coincidence that the earliest date in the next surviving (and fully datable) copy is 1671.[5]

---

4. This is clearly visible in the reduced illustration in the auction sale catalogue on which this description is based.

5. See above, p.20.

It was not, in fact, until 29 August 1678 that Peter Lambecius, Librarian of the Hofbibliothek at Vienna, paid the 400 reichsthalers for the copy 'sent with all loyalty', and noted that it was beautifully coloured, the work of Magdalena Fürstin over five years. Magdalena Fürstin was a person of some distinction, worthy of an individual biographical notice in Doppelmayr's account of the lives of Nürnberg 'mathematici' and artists, as follows.[6]

She was born in 1652, the daughter of Paulus Fürst, bookseller and art-dealer of Nürnberg, and from her earliest years she wanted above all to learn how to paint and draw. She got her wish, since she found instructors in Johann Thomas Fischer and the famous Maria Sibylla Merian, and profited from this opportunity, and by further industry attained a laudable skill herself. The most outstanding example of her work was a copy of the great botanical work on the plants that formerly grew in the garden of the Bishop of Eichstätt, engraved on copper, which she completed in 1677, which work was bought soon after by the Imperial Librarian, the famous Peter Lambecius (and, notes Doppelmayr, another copy of this work illuminated by this artist is in the Stadtbibliothek at Nürnberg). She also executed many small pictures of flowers in water-colour. In her last years she bestowed her talents on a heraldic work, 'das grosse Siebmacherische Wappen-buch',[7] which her father and her second husband, Rudolph Johann Hellmers, had illuminated and sold to the Royal Library, Berlin. In 1717 she was widowed for the second time, and went to live with friends in Vienna, but had been there only a few weeks when she was taken ill herself and died, aged 65.

Magdalena Fürstin's masterpiece is still in what is now the Österreichische Nationalbibliothek. It is sumptuously bound in red morocco. The titlepage has the same novel appearance as BNP, MC and MHN. The background is sky-blue with a broad golden border (similar in colour to MC); the title-cloth is crimson with a green reverse (the latter reminiscent of T, much earlier — but where was T now?). Besler's name and the date are painted out, and in their place is a curious inscription recording that Magdalena Fürstin had clothed it alternately with colours and shading at Nürnberg. The coloured plates themselves are very much closer, in all the details noted, to the 'early' group than the 'French' copies or S. There is one instance of transposition, but this could well be an accident.

But to us the most interesting feature of this copy is the fact that, like the earlier products of the Mack workshop, the plates in this copy are occasionally signed. What is more, they are not exclusively signed by Magdalena Fürstin. The initials 'H.S.F.' and monogram 'HF' are also found, and these make it clear that she had a collaborator in her old teacher, Johann Thomas Fischer, who also rates a notice in Doppelmayr.[8] He was, we learn, born on 21 December 1603, the son of Valentin Thomas, 'Granatrosensetzer in Wöhrd' (Fürth ?), outside the walls of Nürnberg. In his youth he spent a long time in the service of 'eines Junkers Schedel', a member of the famous Nürnberg family. His master, like many distinguished persons in Nürnberg at the time, was a keen friend of painting and had himself learnt how to apply paints in the colouring of engravings and so on. Fischer learnt from his master the art of illuminating 'mit Gummifarben' (colours with a gum arabic vehicle?), and used it to great effect, working with his left hand, to produce renderings of flowers, tulips, roses and carnations. Since he had had no regular training as a 'Briefmaler und Illuminist' (an illuminator of official documents), he could find no regular employment, nor could he take on pupils. He had only one keen pupil, his daughter Anna Katharina, whose married name was Block. He died on 16 October 1685.

The last statement about his life is, as we have seen, not wholly true. But what is particularly fascinating is the knowledge that Fischer learned his irregular trade from one Schedel who must surely have been Sebastian Schedel. There were many Schedels; to be interested in the arts may have been common; but the irregularity of Fischer's subsequent life can only have come from one source. If so, Fischer would have been in his late twenties when both Schedel and Besler died, and could indeed have worked as an apprentice on some of the 'early' copies of the *Hortus Eystettensis*. He would certainly have known about its progress, and may well have kept in touch with the younger members of the Besler family. As a professional flower painter, he would have been the first to see the opportunity in whatever *Nachlass* still existed when Joachim Hieronymus Besler died in 1671.

6. Doppelmayr, *Historische Nachricht von den Nürnbergischen Mathematicis und Künstlern*, 270–1.

7. *J. Siebmachers grosses und allgemeines Wappenbuch*, Nürnberg, 1655.

8. Doppelmayr, 240; Nagler, *Monogrammisten* III, no. 1588.

The evidence suggests that this consisted of one complete copy of the sheets of the special issue on watermarked paper, with all the prelims and the perfected printed text to match; a copy of the plain issue, already bound; and a mass of further sheets, also of the plain issue, mostly those rejected for use earlier, out of which two or more copies might be made up. It may well be that this *Nachlass* was bought by Paulus Fürst or Rudolph Hellmers or both, but there can be no doubt that Fischer and Magdalena Fürstin were to try their hands on it. It is likely that the residue of the original colour guides (whatever they were) may have been included, since the colours are so remarkably close to the 'early' copies. The first and chief element in this, the complete copy, was clearly turned to great effect. Fischer's first dated plate, the 'Colocasia' (plate 346), was a fine test of skill with its huge leaves and delicately veined pumpkins. Perhaps this represents the beginning of the enterprise; the other signatures are spread fairly evenly, though there are surprisingly few among the spring flowers. Magdalena Fürstin's signature in full and the date 1677 on the 'Ficus Indica' (plate 359) may represent the end of this part of the enterprise.

The copy now in the Bayerische Staatsbibliothek (BSB) seems to have been the next (and perhaps last) enterprise in which master and pupil were directly involved. The binding is clearly older than the colouring; whether it once contained more leaves (it ends oddly 35 leaves short of the end of 'Verna') cannot now be discerned. Fischer and Fürstin's work is not happy; despite his skill with 'Gummifarben' the colour spread, even though quickly applied. They seem to have left it incomplete.

Given adequate material, Fischer could produce first-rate work, of which O is not the only example. There is a fine large engraving of the Holy Family with Saint Barbara and Saint Catherine and angels by Stradanus and R. Sadeler in the Germanisches National museum, Nürnberg (Graphische Sammlung, Nor. 379), signed by him and dated '1678 Æ 75½', which shows equally that age had not diminished his skill. Nagler lists other prints coloured by him, including one signed 'Johann Thomas Fischer in seinem 82 Jahre'.

It is interesting that Magdalena Fürstin was not the only woman to provide the local Council with a coloured copy of the *Hortus Eystettensis*. At about the same time, Dorothea Gräfin, wife of Gottfried Graf, Burgermeister of Leipzig, coloured a copy which was kept in the Ratsbibliothek in 1715. No such copy can be traced in Leipzig today, and it seems all too probable that it did not survive the war.[9]

Two other copies, those at Leiden University (LU) and the Bibliothèque Royale, Brussels (BR), may stem from the Besler *Nachlass*, although the colouring technique, similar in both copies, has nothing in common with the work of Fischer and Fürstin. Perhaps when Fischer's technique and colours proved unequal to the inadequate paper, they or the owners of sheets decided to cut their losses and disposed of them. The new technique involved the exclusive use of water-based colours, thinly and fairly evenly applied so that the engraved detail alone is used to provide shading, without the use of a darker colour. Fischer would no doubt have regarded it as an unsatisfactory technique; it certainly demands more of the viewer, being not so much a 'counterfeit' of nature as an allusion to it. It is interesting that both copies seem to have come to Holland at about the same time. The Dutch taste for fine engravings of flowers had inspired some of the original engravers; it now seems to have absorbed the last remnant of their work.

LU is particularly interesting because it preserves a fragment of the 'early' style of colouring, the unfinished but beautiful season-title for 'Verna'. Whether this represents Fischer's unfinished work, or was part of the *Nachlass* and therefore still earlier, it is impossible to say; it is not like any of the other season-titles, even among the 'early' group. The title-pages in both LU and BR are perfunctorily rendered, if with some knowledge of the tradition. The plates in both have a homogeneous appearance, in both increased by the generally dark colour of the prints. Evidently, these are sheets from the end of the original run, when the engraved detail was becoming full of partially dry ink and increasingly difficult to wipe clean. The colour variants are interesting; some of the differences correspond with the older tradition, although the majority are 'late'. The most striking feature, however, is the transposition of colour in two different ways in the two copies. Although not so frequent a feature as in S, it is clear evidence that both copies were coloured, at least in part, from a guide, a sketch with verbal description, rather than a coloured exemplar.

---

9. 'Amaranthes' [Gottlieb Siegmund Corvinus], *Nutzbares, galantes und curioses Frauenzimmer-lexicon* (Leipzig, 1715), cols 678–9.

Inevitably, some copies lie outside the groupings hitherto suggested. One such is the fragmentary copy at Ellwangen (Ellw). To all appearances it is a conventional 'plain' copy, in the sort of binding produced with remarkably little change in Germany throughout the seventeenth and into the eighteenth century. This one appears to have been made about the middle of this period, certainly well after the book was printed. The colouring, too, might at first sight be thought amateur work, painted by a local artist from nature, but it is not: despite the fact that only a few sheets are coloured, and most of those only partially, it is clear that the colouring was taken from an exemplar of the 'early' type, which is followed with professional exactitude. Even more interesting is the 'faded' green used for the foliage. Despite the sad misadventures to which the book has been subjected in recent and perhaps earlier times, there is no suggestion that this is the result of subsequent damage. The tint closely resembles that used for similar foliage in the 'plate' drawings (ErZ), as copied by or for Trew in an otherwise plain copy of the original edition[10]; another point of resemblance is the colouring of details only. It may be that there is some further connection between the two, so far elusive.

There may be, certainly were, other coloured copies of the original edition of Besler, but those described (and listed below) are all I have been able to find. The loss of the copy coloured by Dorothea Gräfin and once at Leipzig is particularly sad. There remain two more copies to consider, those of the '1713' edition, in fact printed $c.1750$. Only one of these is complete, that in the Oak Spring Garden Library, Virginia (OS). Its binding is more or less contemporary with the printing. The title-page recalls that of O, and the arms of Bishop Johann Anton I von Katzenellenbogen have been correctly rendered. There are some signs that the artist was familiar with the 'later' tradition of the original edition, in particular a curious error shared with MC. On the other hand, there are many more signs that an entirely new source (or sources) was available. Some of the new features must be based on drawing from nature, seen with a new, eighteenth-century eye, conditioned by the scientific discoveries of the last 50 years, notably the work of Linnaeus. The volume of botanical writing, much of it illustrated, had greatly increased, and with it knowledge of the anatomical structure and habit of plants. Many that were exotic new imports in 1613 were now thoroughly acclimatized. Some of this can be seen in the colouring of the Oak Spring copy, restricted though the artist is by the engraved image now almost a century and a half old. The style of painting, too, reflects the new taste for 'watercolour', translucent tints based on a wide variety of pigments, many newly available. These are applied with great skill, given the unsuitable nature of the paper.

It is puzzling that there is no reference to such a copy, if strictly contemporary, in Trew's correspondence (so far as it is catalogued), but neither is there any reference to his own acquisition of the 'plate' drawings. The fact that these are now bound with the gratulatory sheet addressed by Besler to the Collegium Medicum in 1627 and a print of Gärtner's 'Zierblatt' suggests that Trew may have picked this up in Nürnberg, now the forgotten residue of a Besler connection or of one of the colourists. The presence of the titlepage, as well as the other printed sheets, preserved and bound with an archivist's care by Trew, suggests the latter. This is confirmed by another of Trew's acquisitions, an imperfect set of text sheets, almost all printed on both sides, as provided for coloured copies; perhaps this was acquired by Trew from the same source.[11]

The copy at the Pfalzische Landesbibliothek at Speyer (Sp) is, like that at Ellwangen, incomplete. Again, its appearance at first suggests amateur work, but the fact that the plates coloured come close together, and are not spread haphazard as they would be if taken from nature, makes it clear that here again a model of some sort was followed. The artist was of some ability, defeated, perhaps, by the unsympathetic paper.

The colouring of the *Hortus Eystettensis* forms part of the development of the depiction of nature over two centuries, from the late sixteenth to the eighteenth century. In it can be seen the reaction of artists, mainly professional artists, to the demands made by other professionals, at first physicians and apothecaries, but, increasingly, professional botanists, scientists to whom the understanding of nature now provided a livelihood. The knowledge of nature had

---

10. Now University Library, Erlangen, Trew B1.
11. The drawings are in an elegant pale calf binding, clearly made for Trew; the fragment is now Trew B 5.

all the freshness of novelty when Joachim Camerarius, Abbot Johann V Müller and Bishop Johann Conrad von Gemmingen first studied and deployed it in their different ways. But the *Hortus Eystettensis* was not large because it contained many new plants: it was a 'summa summarum', a deliberate attempt to comprehend, as well as depict (sometimes more than once), the whole of the vegetable creation, recorded as well as grown on the Willibaldsburg. It was this that gave it such immediate appeal when it first appeared, and ensured its survival long after the garden, in the form in which the Bishop had conceived it, had ceased to exist. Indeed the ravages of war gave it a special, almost pre-Adamite, fascination as a picture of paradise before the Fall.

All this was bound up in the attempts by successive generations of artists to give colour to the images rendered in black and white with such conspicuous success by Wolfgang Kilian and his contemporaries. It was, in a sense, always a losing battle. Kilian or whoever drew the 'plate' drawings thought in black and white: the engravings plain are themselves a sufficient rendering, as far as botanical detail and light and shade go. Colour, if an extra grace, was also supererogatory: the artist always had to find a way to add colour without drowning or being drowned by the engraved detail. This could be done with conspicuous success, and DeB, on the one hand, and the Mack copies, on the other, represent two different such attempts, one essentially modern and 'realistic' in its approach to colour, the other employing the traditional skills of the miniaturist. Other attempts were less successful, due to the intransigence of the printed image or of the paper, or simple lack of artistic talent.

As time went on, the expectations of the human eye changed again. Instead of having a solid, opaque, almost three-dimensional, 'counterfeit' of nature, colour became a sort of stage gauze, a veil through which you could see the reality of nature. This change was related to the improvement of optical instruments like the telescope, with its tendency to flatten objects seen at a distance: nature became less three-dimensional and more a series of two-dimensional images. Botanical drawing in the eighteenth century became lighter and more impressionistic, at the same time as detail (visible more clearly than ever before through the microscope) became finer. The two merged in the shimmering minuteness of Ehret or Redouté.

By this time the powerful baroque images conceived by Camerarius and brought to fruition in the *Hortus Eystettensis* had become history. But that history still has power to move us today. With eyes more attuned now to the power and density of, for example, Rubens (who owned a copy of the *Hortus Eystettensis*),[12] we can appreciate the strength, as well as the decorative skill, whether in the detail of the engravings or transformed by colour, of the images in the *Hortus Eystettensis*, from the minute heartsease to the great Agave or 'Ficus Indica'. Their power to fascinate today is demonstrated by the success of the several facsimile editions. But nothing, except the originals themselves, can quite communicate the majesty of the coloured copies of the *Hortus Eystettensis*.

12. H. G. Evers, *Peter Paul Rubens* (Munich, 1942), 30.

CATALOGUE OF THE COLOURED COPIES

K

LOCATION: Library and Archives, Royal Botanic Gardens, Kew, 'Schedel Calendarium'.

BINDING: red morocco over pasteboards, gilt roll-tooled octagonal panel, within roll-tooled gilt border, arabesque inner corners; rebacked and re-endpapered c.1900, original book-plate re-laid down.

CONTENTS: 281ff, foliated 1–289 by Schedel, but with some leaves numbered from a different earlier sequence, others inserted and others now missing. Full contents listed (Appendix D).

PAPER: Reçute folio; watermarks: Arms of Nürnberg (e.g., f.12), Arms of Pfalz-Neuburg (e.g. f.16, cf. Briquet 1970–81 but with moor's head on an escutcheon below), Castle watermark (f.263).

COLOUR: f.50v: 67iv and v, f.51v: 67iii, f.52v: 67ii, f.54: 67i, f.109v: 70i, f.112v: 70iii, f.117Av: 70v, f.118Av: 70iv, f.160: 290ii, f.160v: 123ii, f.161: 123i and ii, f.161v: 123iii, f.162: 123i. These fragments, clearly contemporary, were perhaps discarded during the process of colouring another copy. The range of colours, and tone (darker), suggests one closer to DeB than to the Mack copies. There is no trace of pink or red on 70 iii, and the blue centre is an improbable pale turquoise, which may explain why it was discarded. See pp. 31–32 above.

OTHER FEATURES: For drawings and text, and their resemblance to the *Hortus Eystettensis*, see Appendix D below.

PROVENANCE: Sebastian Schedel (1570–1628), on whom see pp. 31–3 above; acquired after the death of Sir Joseph Hooker (1785–1865) but before 1890 by the Royal Botanical Gardens.

ErZ

LOCATION: Erlangen, Universitätsbibliothek, MS.2370.

BINDING: One volume, brown calf over paste-boards, plain gold fillet border, gilt spine, blue sprinkled edges, $18C^2$.

PRELIMS: Title, dedication IOHANNI CUNRADO, 'Decanus . . . lectori' and 'Ad nobilem . . .', 'Inclutae Reipub.', portrait and arms/blank, 'Zierblatt' by Georg Gärtner.

CONTENTS: 'Plate' drawings, pen and grey wash, for all 367 plates, (except as noted below), margins trimmed and mounted on sheets of 18C paper. No foliation.

PAPER: Royal broadside, most of the sheets without watermark, but the main grapes watermark is found (e.g. f.107), and also Arms of Nürnberg (e.g. f.6). The fact that the same type of paper is used here and in the watermarked printed copies suggests that the various paper stocks were all available before printing began and do not constitute evidence of the priority of any of the different issues.

COLOUR: The printed pages are uncoloured. On the drawings, coloured details occur only sporadically as far as plate 133, with quite precise colour descriptions of individual parts (e.g. 'licht grau grun', plate 16) in a small neat hand. The tulips (plates 67–75, 77–8) are extensively coloured, no doubt due to the difficulty of indicating colour by detail only.

SPECIAL FEATURES: The names of the plants are given in exactly the same form as on the printed plates, written in a neat large upright hand, with no distinction between roman and italic; the names of the tulips are in a still more formal script, imitating roman type. The 'transfer system' marks (see p. 30–1 above) are clearly visible up to f.81; those on f.79 (plate 67) are, unusually, numeric not diagrammatic. One sheet, f.92 (plate 80), is clearly a replacement, presumably copied in reverse from the print, made by or for Trew, on the same paper as that used to mount the original drawings; f.100 (plate 88) may also be a replacement (the annotation is in a very different hand). The design for 'Asphodelus Liliaceus' (plate 130 iii) has been left in pencil and not completed. There is a smear of ink from the edge of a plate on f.44 (plate 32), which shows that the drawings continued in use by the print-makers after the plates were engraved. The sheets were also originally numbered, perhaps by the same hand that wrote the captions, in an order that bears no relation to the present sequence. In some cases the number has been corrected; thus, the present plate 1 is numbered '13' and '236', and 15 is corrected from '30' to '22'.

PROVENANCE: Probably retained by the Besler family, and possibly used as a partial guide to later colouring (see p.31); acquired by Christoph Jakob Trew after 1752 but before c.1765, the dates of, respectively, his printed *Librorum botanicorum libri duo* and subsequent manuscript catalogue (Erlangen, Universitätsbibliothek, MS.2474), in which it is recorded for the first time as 'Horti Eystettensis primae atramento Sinico repraesentationes cum aliquot singularum partium coloribus nativis, fol. max.' Trew also caused the few coloured details to be copied into his other, otherwise uncoloured, copy of the *Hortus Eystettensis* 1613 (Trew B1). Bequeathed by Trew to the University of Altdorf (1769), and transferred to Erlangen on its closure in 1818.

N

LOCATION: Nürnberg, Stadtbibliothek, Solg. 2°. 1797.

BINDING: Two volumes, brown hide over wooden boards, bevelled on inside edge, blind tooled with central arabesque lozenge plaque, enclosed by a vine-wreath roll border within fillets and by double lozenge compartments made with a 'thunderbolt' roll within fillets, imposed on a similar rectangular double panel, within outer border, roll gilt within fillets, gilt and gauffered edges, plain end papers, $17C^1$.

PRELIMS: Title, dedication IOHANNI CUNRADO, 'Decanus . . . lectori' and 'Ad nobilem', Belgian/Dutch and French privileges, 'Autores'/portrait and arms.

SEASON TITLES: All four, with 'Aestiva' as title to second volume.

CONTENTS: Plates alternately on verso and recto, interleaved with printed text, backed up so as to face relevant plate, 'Verna' followed by 'Autumnalis' and 'Hyberna', 'Aestiva' in second volume. Indexes follow parts. No foliation.

PAPER: Plain laid, 32mm between chain lines, varying thickness.

COLOUR: Title, background dark ultramarine with broad black border, title cloth crimson with pale yellow simulating shot-silk, lettering and fringe outlined in gold throughout, reverse dark red: signed 'Geörg Schneider/pinxit'; portrait and arms on dark ultramarine panel, signed 'Geörg Schneider/pinx. 1613; 'Aestiva' part-title signed 'Georg Schneider Illuminist. N: 1614'. The plates are extensively signed with initials or names of Schneider, George Mack, 'H.S.', 'M.B.' and 'D.R.' At least one other artist was involved who did not sign his name. Several plates (e.g. Plate 222) have a marked fine outline which does not appear in the work of the named artists. Plate 120 is dated '1613' by Mack, and Plate 359 by Schneider 'a° 1613 dies Maiūs · 4'. There is a close resemblance, in style and palette, and in the incidence of signatures, to T.

OTHER FEATURES: Old (18C?) armorial bookplate of 'Bib. Nor.' on title; old, perhaps contemporary, inscription on label on front pastedown 'Hortus Aistettensis/Horti delicias qui non amat, exeat Horto/ Horti epulas omnes, Hortus hic unq' fui/C.R.' [?G.R.]. Plate 4 has offset of another impression in inner margin; plates 23, 35–7, 79, 84, 89, 97, 99, 145, 194, 234, 274, 332, 356 and 360 have traces of oil and folding, and were probably removed for tracing and replaced; plates 242–3, 282, 300 and 302 have duller colour and less detail.

PROVENANCE: Although shelved by the Stadtbibliothek as part of the collection of Adam Rudolph Solger (1693–1770), acquired in 1766, this copy already belonged to the City. Although a copy is listed in the printed catalogue of Solger's collection (*Bibliotheca, sive supellex librorum impressorum . . .*, Nürnberg, 1760–2, I. 168), the entry is annotated by the then librarian: 'Dieser hier aufbewahrte Exemplar fehlt: es wurde als Dublette in älteres Zeit verkauft. Ich habe um die Nummer auszufullen ein anderes Exemplar welches die Stadtbibliothek besaß hier ein gereiht zwei Bände unter (2). 1797.a.b. Sie stehen wegen ohrer Höhe auf dem letzten Gestell in Solger Complex.' This may, therefore, be the copy requested from Besler by the Rat, and supplied between 22 July 1613 and 1617 (see p.36–7 above). The distich is perhaps (like the lines on the engraved portrait of Besler) by Georg Rem.

T

LOCATION: Turin, private collection.

BINDING: One volume, dark brown calf over pasteboard, border of broad scrolled floral roll enclosed by double gilt fillets, outer arabesque roll and inner dog-tooth roll, joined by roll of crossed flowers (pansies?) to outer panel of arabesque and dog-tooth rolls with double filler between, large arabesque inner corners, enclosing inner panel of double fillet, outer flora cornerpieces, large double volute inner corners, large floral vesica-shaped centrepiece with small oval centre, all gilt; spine with raised bands, compartments with gilt centres and corners, title lettering piece; marbled endpapers, gilt edges, 17C².

PRELIMS: Title, dedication IOHANNI CUNRADO, 'Decanus . . . Lectori' and 'Ad nobilem . . .', Belgian/Dutch and French privileges, portrait/and arms/'Autores'.

SEASON TITLES: None.

CONTENTS: Plates for Spring and Autumn only, alternatively on verso and recto, interleaved with printed text, backed up so as to face relevant plate.

PAPER: Plain laid, 32mm between chain lines, varying thickness.

COLOUR: Title, exactly as N, except that there is a thin red line within the black border, the reverse of the title cloth is dark green, the fluting on the pillars is gold not silver, the 'D' of 'Diligens' has a dark red background instead of black (but identical gold outlining and hatching) and the colour scheme of the mantling at the base of the coat of arms is different; it is signed 'Georg Schneider 1613', in the same place, below the pediment of the left-hand pot; portrait and arms signed as N. The plates are signed by most of the same artists as N (Schneider, Georg Mack, 'H.S.' and 'D.R.'), usually (but not always) in the same place and form. Plate 120 is dated '1613' by Mack, and plate 359 by Schneider 'A°. 1613/dies Maiūs. 4.'

OTHER FEATURES: The title-leaf has been cut to the edges of the painted black border and laid down on white laid paper, probably contemporary with binding. Below the title is '52b(?) duplum' in a 17C² or 18C¹ hand. The portrait leaf and first 20 plates have been mounted on guards, perhaps the result of earlier damage; the portrait leaf and plates 10, 22, 23, 27, 35, 38 and 133 are discoloured, perhaps also the result of damage or tracing. The title 'Sextus Ordo Vernalium' has been cut out of another sheet and pasted to the verso of plate 94. Foliated in pencil, in a German hand, 20C¹.

PROVENANCE: Possibly one volume of the copy supplied by Besler and sent to Eichstätt, presumably for the Bishop and Chapter, which cost 500fl., according to the contract of 23 September 1613. If so, it may have been looted during the sack of the Willibaldsburg in 1633–4, and this part separated from its counterpart (see L, below); the present binding suggests that it remained in Germany, its original binding perhaps so damaged as to require replacement. It may at this stage have been in an institutional library, hence the inscription on the title-page. It was later in the collection of David Samuel von Madai (1709–80), with his bookplate. Madai was originally from Schemnitz (Hungary) but studied at Halle, where he succeeded his father-in-law as 'Arzt am Waisenhause'. He became court physician to the Fürst von Anhalt-Cöthen in 1740. He published a number of medical books, but was better known as a coin-collector, the author of *Vollständiges Thaler-Cabinet* (Königsberg, 1765–7). Recently in the German booktrade, until bought by the present owner.

L

LOCATION: Leufsta, Uppland, Sweden (Uppsala University Library).

BINDING: One volume, brown hide over paste board, blind tooled with central arabesque lozenge plaque in large lozenge compartment made with a 'thunderbolt' roll within treble fillets, imposed on double fillet panel, arabesque floral outer and inner corner pieces, all within outer border of a broad floral roll enclosed by vine-wreath rolls, each enclosed by treble fillet; spine with seven double raised bands, small arabesque lozenge centrepiece with central void in compartments; four clasps (hasps only remaining on upper board); gilt edges; $17C^1$.

PRELIMS: None.

SEASON TITLES: Aestiva and Hyberna.

CONTENTS: Plates for Summer and Winter only, alternately on verso and recto, interleaved with printed text, backed up so as to face relevant plate. Foliated in ink by Olaf Rudbeck the elder (see below).

PAPER: Plain laid, 32mm between chain lines.

COLOUR: Plates signed by 'G.S.' (Georg Schneider), Georg Mack ('G.M.', 'G.M. Seni.'), 'G.H.' and 'D.H.' 'Classis Hyberna' signed and dated 'Georg Mack 1614'.

OTHER FEATURES: Plain endpapers, numbered '153' on front pastedown; foliated in ink, as above; the binding and first leaf slightly damaged.

PROVENANCE: The binding resembles that of N in design, though not in detail; two of the ornamental rolls used in both appear to be identical. Possibly, therefore, this is the second volume of a set of which T is the first, although there is not the same striking correspondence in the colourists' signatures. If so, then part of the set sent to Eichstätt by Besler, and looted by Swedish troops in 1633–4. The number '153' written on the front pastedown of this and its companion volume (uncoloured and in a different binding) indicate common ownership of both volumes before they came to their next owner. Next it was acquired by Olof Rudbeck the elder (1630–1702), by whom left to his son Olof Rudbeck the younger (1660–1740). Slightly damaged in the Uppsala fire in 1702. Bought at the sale of Olof Rudbeck the younger's widow by Charles de Geer (1720–78) of Leufsta, where it has been ever since. The de Geer collection was acquired by Uppsala University between 1985 and 1989, jointly by purchase by the Swedish government and through a donation by Katarina Craaford, but remains *in situ* at Leufsta. See Tomas Anfält, 'A Consumer of Englightenment: Charles de Geer, Savant and Book Collector in Eighteenth-Century Sweden', *The Book Collector*, Summer 1991, 197–210.

BL

LOCATION: London, British Library, 10 Tab. 29.

BINDING: Brown russia/morocco over pasteboards, broad thick and thin fillet border, arms of George III in centre, gilt; flat spine gilt in compartments; blue marbled edges and end-papers. $18C^2$.

PRELIMS: Title, dedication IOHANNI CHRISTOPHERO, ['Decanus ... lectori' and 'Ad nobilem', French and Belgian/Dutch privileges,] blank recto/portrait and arms.

SEASON TITLES: None.

CONTENTS: Plates on versos throughout, with German plant names written in a neat large textura on the facing recto, in the order 'Verna', 'Aestivalis', 'Autumnalis', 'Hyberna', followed by the indexes bound together in the same order. No foliation.

PAPER: 'Grapes' watermark, 30mm chain lines, except for portrait and arms leaf (plain, 32mm chain-lines).

COLOUR: Title, background pale sky blue, title cloth pale yellow, reverse dark red, detail plain black: signed 'Georgius Mack/Illuminist: in Nurnberg/Anno 1615/Aprilis 16' across vase pediments. Dedication initial painted. Portrait and arms on sky blue background, signed 'Geo. Mack 1614', and elaborate scroll-work round the letter-press part-titles. The plates are signed by Mack almost throughout, plate 78 dated '2 Marti. 1614', plates 80, 86, 87, 109, 114, 125, 127, 134, 187, 212, 213, 221, 225–7, 259, 272, 280, 400, 309–10, 317, 334–7, 345, 348, 358 & 361–2 all dated '1614'. Although largely conforming to the colour pattern of other early copies, BL has some distinctive features, e.g. the use of a specific lilac tint (e.g. Plates 16 ii, 167 i. 170 ii, and 232 ii). The frequent use of a contrasting base colour with a secondary tint, applied with very fine strokes, is also peculiar to this copy. The treatment of *Iasminum Indicum* (plate 335 i) is not found elsewhere. See App. C.

SPECIAL FEATURES: Although now bound in one volume, the non-sequential order of the seasons suggests that it may have once been bound in two volumes. It was re-trimmed when bound in the 18th century, but some leaves could not be cut since there was too little margin; these were folded in and the edges show that they were orginally gilt, like N and L. The leaves are double-ruled in dark red, except those for plates 97 and 316. Plate 182, once mounted at the head edge of plate 181 (of which it forms the upper part) is missing; plate 102 is mis-bound between plates 88 and 89. The German names on rectos, though written with professional assurance, do not always fit neatly; only Er has similar captions, sporadically.

PROVENANCE: This is Georg Mack's masterpiece and, if the signatures are to be believed, almost all his own work. It must have been executed for some major client. It is tempting to think of Wilhelm V of Bavaria (1548–1626), who expressed such interest in Bishop Johann Conrad's collections and garden, and at whose behest Hainhofer made his journey and the report on it that followed, or his son Maximilian (1573–1651). A number of copies (besides the three now remaining) of the *Hortus Eystettensis* have been lost from what is now the Bayerische Staatsbibliothek, and it may be that even this copy, perhaps in an old-fashioned and shabby binding, was considered superfluous. It had entered George III's possession before 1780, to judge by the binding, and passed to the British Museum, and thence to the British Library, with the rest of his library at the behest of George IV in 1824.

B1

LOCATION: Berlin, Staatsbibliothek, Libri Picturati A.163 (transferred in 1968 from general collection; previous shelf-mark Lz 25910ª R).

BINDING: One volume, half dark brown morocco, papersides, fine red and blue sprinkled edges contemporary with present binding, 19C^2.

PRELIMS: Title, dedication IOHANNI CUNRADO, 'Decanus . . . lectori', 'Ad nobilem . . .', French and Belgian/Dutch privileges, indexes ('Verna' — 'Hyberna'), portrait and arms/blank.

SEASON TITLES: None.

CONTENTS: Plates on versos throughout, with text (a condensed version of printed text) written on facing rectos, the German names in neat fraktur, the rest in roman script. Foliated in the same hand.

PAPER: Plain laid, ±32mm between chain lines.

COLOUR: Frontispiece, background dull slate blue with broad black border, title cloth yellow, reverse bright blue, detail plain black; initials and tail-pieces in prelims all painted; no background to portrait and arms. Elaborate scrollwork round the letterpress part-titles. Early plates rather perfunctorily coloured, except for plate 66, signed by Georg Mack; thereafter signed regularly, and plate 169 dated '1615'. The detail of the plates mostly conforms with that of other early copies, with three unique discrepancies at 65 i, 83 ii and 232 i.

SPECIAL FEATURES: The manuscript hand occasionally corrects the plate captions, e.g. substituting 'martia' for 'matronalis' at 18 iii. Plate 44 is dirty; plates 81, 190, and 222–3 are missing. Double-ruled in red throughout. At the beginning of the book, before the frontispiece, there is a manuscript 'Index Horti Eystettensis accuratior, et universalis CIƆIƆLIX' on 8 leaves. The date is a puzzle: 1659 is perhaps more likely than 1609; it is certainly 17C, earlier than the much larger manuscript 'Index Horti Eystettensis' at Erlangen (Universitätsbibliothek, MS.890).

PROVENANCE: Possibly acquired by the Prussian Royal Library from Ansbach (see above, p.40); 19C^1 stamp 'Ex Bibliotheca Regia', and thence to the Deutsche Staatsbibliothek.

V

LOCATION: Rome, Bibliotheca Apostolica Vaticana, Barberini XI. 16–17.

BINDING: Two volumes, red morocco, central cartouche (centre now cut out, once with Barberini arms?), enclosed in hexagonal roll-tooled panel within rectangular panel, arabesque roll enclosed in fillets, arabesque fringe roll repeated on either side; border of broad floral roll with in double fillets, arabesque fringe roll repeated on inner edge, all gilt; seven raised bands, gilt compartments. 17C^2. (The binding has been 'improved' with elaborate rococo tooling round the central cartouche and similar cornerpieces between border and panel, 18C^2.)

PRELIMS: Title, dedication IOHANNI CUNRADO, 'Decanus . . . lectori', French and Belgian/Dutch privileges, 'Autores'/ portrait and arms.

SEASON TITLES: All four present, uniformly coloured (centres shaded crimson).

CONTENTS: Plates alternately on versos and rectos throughout, printed text interleaved throughout, backed up so as to face relevant plate, 'Verna' in volume 1, 'Aestiva', 'Autumnalis', and 'Hyberna' in volume 2; indexes bound in after each season. Remains of foliation in 17th-century Italian hand [1]–[428], mostly trimmed away.

PAPER: Plain laid, 30 mm between chain-lines.

COLOUR: Title, background dull sky blue, black border, title cloth yellow, reverse strong blue-green, detail plain black; first initial not coloured, but others, including tail-pieces, painted in; no background to portrait and arms (the flowers on the sprig of basil held by Besler is red). The plates are sparsely signed with initials by Georg Mack ('G.M.', 'G.M.A.'), with apparently unfinished signature and date (4 June 1615?) on plate 78; other initials 'GAH' (plate 215) and 'M^9' (plate 245) are probably also his. The colour pattern conforms to the conesensus of early copies, without exception.

SPECIAL FEATURES: Double-ruled in red throughout.

PROVENANCE: Old shelf-mark 'XXIII-F.4–' 'F2–' on title; 'In Ind[ic]e vide – Eystadiens' at foot of title. Possibly reached Rome Bibliotheca Palatina in or soon after 1622; bound for Cardinal Francesco with the Barberini, Vatican Librarian from 1626, and reached the Vatican Library with the Bibliotheca Barberini in 1902. See Jeanne Bignami Odier, *La bibliothèque Vaticane de Sixte IV à Pie XI* (Studi e Testi 272), Citta del Vaticano, 1973, 109, 242, 255n.

BNM

LOCATION: Madrid, Bibliotheca Nacional, Servicio de Dibujos y Grabados, ER 1216–17.

BINDING: Recent half dark brown morocco with cloth sides, new endpapers, old plain edges.

PRELIMS: Title, dedication IOHANNI CUNRADO, 'Decanus . . . lectori', 'Ad nobilem', French and Belgian/Dutch privileges, portrait and arms/'Autores'.

SEASON TITLES: All four present, coloured (see below), 'Verna' bound after trees and before flowers.

CONTENTS: Plates alternately on versos and rectos throughout, printed text interleaved throughout, backed up so as to face relevant plate, in season order; indexes bound in after each season.

PAPER: Plain laid (32mm between chain-lines), variously thick and thin.

COLOUR: Title, background dark brown with thick (5mm) black border, title cloth dark yellow reverse crimson highlighted with gold, main lines of title outlined in gold (including 'Solomon' but not 'Cyrus'); prelims plain, except portrait and arms (correct), which have white background with thick black border (basil flowers crimson), initials beneath and on verso coloured. The season titles are strongly and diversely coloured; the 'Verna' and 'Aestiva' panels are green,

'Autumnalis' yellow and 'Hyberna' grey. The colouring of the plates conforms to the consensus of early copies, with idiosyncrasies, some of which (like the brown title background) it shares with Eich. The plates are sparsely signed, some by Mack ('GM' 93, 'IM' 117); other initials 'IN' (146, 153), 'R' (183) and 'RTAF' (336).

SPECIAL FEATURES: The 'Verna' index, text for 'Quartus ordo autumnalis' (plates 354–60), plate 360 and text for plates 363–4 are missing. Plates 191 and 213 have not been wiped clean, plate 359 is damaged at head, and plate 361 is browned, probably from contact with oiled tracing paper.

PROVENANCE: Clearly of German origin, this copy may have come to Spain after 1665, with the Austrian regency, or the change of dynasty in 1700; it is not listed in the 1637 'Indice de los libros que tiene su magestad', but is present in the Biblioteca Real catalogue of 1746 (B.N. MS 18828); small 18th-century bookplate and stamps of Biblioteca Real.

W

LOCATION: Wiesbaden, Hessische Landesbibliothek, Rara H. Gr 2° Sh. 7846.

BINDING: One volume, dark brown hide (now virtually black), central flower vase device, surrounded by triple panel, floral cornerpieces, all gilt. 17C^1, much damaged and repaired, two modern clasps.

PRELIMS: Title, dedication IOHANNI CHRISTOPHERO, French and Belgian/Dutch privileges, portrait only (verso blank).

SEASON TITLES: None.

CONTENTS: No text; plates bound as rectos, except where backed up by part-titles, where part-titles appear on recto, with following plate on verso. Later ink foliation.

PAPER: Laid, 'grapes' watermark, 30mm between chain-lines.

COLOUR: Title, slate grey background, broad black border; title cloth pale yellow, shaded darker, detail black, with lettering outlined in gold, dedication initial coloured; the portrait, with green background and yellow inscription, is surrounded by an oblong red panel. The title is signed below the cartouches at the foot of the columns, 'Georg Mack', 'Illum. 1617'. There are no signatures on any of the plates, whose colouring rarely diverges from the consensus of early copies, and then only inconspicuously (e.g. plate 90 iii and iv).

SPECIAL FEATURES: Title substantially repaired, plates 2–8, 10, and 13 missing and remainder to plate 18 bound out of order, plates 86–7 and 95 missing, 103 bound after 108, 106, 154 and 313 missing; manuscript index on 9 leaves at end.

PROVENANCE: Acquired by the Hessische Landesbibliothek by 1867.

B2

LOCATION: Berlin, Staatsbibliothek, gr. 2° Lz 25910 R.

BINDING: One volume, half brown morocco, cloth sides, new white endpapers; older edges, perhaps 18C, sprinkled red (some leaves at end retain still older gold edges); the binding 20C^1, endpapers renewed since.

PRELIMS: Title, portrait and arms/'Autores', dedication IOHANNI CUNRADO, 'Decanus . . . lectori', 'Ad nobilem', French and Belgian/Dutch privileges.

SEASON TITLES: All four present, coloured (see below).

CONTENTS: Letterpress on rectos, plates on versos, in order of seasons.

PAPER: No watermark (chain-lines 32mm).

COLOUR: Title, background pale grey (?oxidised white) enclosed in broad (6·5mm) black border, broader (9·5mm) at the base, title cloth yellow, shaded darker, crimson reverse, detail plain black, portrait and arms without overall panel (background to portrait crimson, arms incorrect with cloud instead of mount), B initial and subsequent initials in prelims all painted. All season titles elaborately painted in different and contrasting colours; the centre panel for 'Verna' is yellow-green, 'Aestiva' yellow, 'Autumnalis' brick red and 'Hyberna' blue-green. In general, the colouring of the plates is rough, hardly keeping to the engraved outline (generally replaced with a dark red painted line); the outlines of the layers of bulbs has been carefully outlined in a darker tint (except plate 111). In detail, B2 conforms closely to the consensus of early copies, but is the first to show red berries at 198 ii and red flowers at 288 i.

SPECIAL FEATURES: The text is preceded by a full size vellum sheet with the elaborate (nine quarterings) painted arms of Philipp II, Duke of Stettin and Pomerania, the date 'A° 1617' and beneath:

Sereniss̄. & illustriss°: Principi Dño PHILIPPO /II Duci Stettin. Pomer. Cassub. et Vandalor, principi Rugiae; /comiti Grutzko, &c. principi et Dño meo clementiss°.

> Mille coloratum Tibi dono floribus Hortum,
>   Inclyte Stetini Dux Pomerane soli
> Verbere languescent ut grandinis: haud tamen his Rex
>   Ornatu Salomon par, similisve fuit.
> Florent namq' homines, sed flos est iste caducus
>   Nil stabile in terris praeter Amare DEVM.
> Sereniss.^{e} Celsit.^{is} Vr̄e/Humillim^s
> Servus/E.F.E.St.F.M./Philippus
> Hainhofer civis Augustan'.

PROVENANCE: Presented (*ut supra*) by Hainhofer to Philipp II, presumably having failed to have a copy presented, or sold, to him. The gift was still vividly recalled by Hainhofer almost thirty years later (see Appendix A, p.65). The Duke (1572–1618) had died untimely, and the dispersal of his estate evokes lament in terms that recall those of Hainhofer's limping verse. His collections passed first to his brother Franz (d.1620), then to Bogislaw XIV (d.1637), with whom the dynasty ended; by then the Thirty Years War was raging, and the collections were dispersed or lost ('nach dem Aussterbendes pommersches Herzogshauses, sind sie unter der Verwirrung des 30jahrigen Kriegs teils zersplittert, teils verschollen': *Allgemeine Deutsche Biographie*, XXVI, 34–6). This copy was in the Prussian Royal Library not later than 1746–7, the date of Johann Carl Wilhelm Möhsen's *Dissertatio epistolica de manuscriptis medicis quae inter codices bibliothecae regiae Berolinensis servantur*, in which it is described (page 64); thence to the Deutsche Staatsbibliothek.

DeB

LOCATION: U.S.A., private collection.

BINDING: Two volumes, tawed hide over thick pasteboards, blind-tooled overall with rolls, remains of two pairs of clasps at foredge, the joints, repaired 19C[2] and new headbands (?) added, with new thin endpapers (these were replaced with stronger acid-free paper and the repairs to the joints made good in the British Library bindery, 1990), original sewing and dark blue edges; C17[1].

PRELIMS: Volume 1: dedication IOHANNI CHRISTOPHERO, French and Belgian/Dutch privileges, 'Verna' season-title. Volume 2: title, 'Aestiva' season-title.

SEASON TITLES: Three present ('Aestiva' is missing and the engraved title bound in its place).

CONTENTS: Plates on rectos only throughout, no printed text except part-titles (on versos); German names written in a contemporary current hand beside the engraved Latin names; contemporary manuscript foliation.

PAPER: Laid, 'grapes' watermark, 30mm between chain-lines; eight leaves (ff.186–193) after 'Aestiva' and endleaves in volume 2 on paper watermarked with a mitre and FS.

COLOUR: Title, dark grey-brown on black background extending to present trimmed edges, title cloth yellow, reverse yellow, shaded dark red. There are no signatures on the plates. The colouring is of exceptional richness and diversity, with a wider range of colours than any of the other early copies. There is a marked absence of gold or any of the other purely ornamental techniques used in the Mack workshop. The botanical accuracy is similarly superior, and on numerous occasions uniquely records the right colour (see below, pp.68–72).

SPECIAL FEATURES: The texts of the German names are longer than those in BL and Er, sometimes with references not included in the printed text. The 'Quartus Ordo' part-title in 'Aestiva' is upside-down, although the plate is the right way up. At the end of the first volume (previously pasted to the 19C endpapers, now removed) are a drawing of the 'Aloe Americana' that flowered in the garden of the Markgraf of Ansbach (see plate 80). With this is the engraving by Wolfgang Kilian, made in 1628 (see p.39), with a long inscription on its flowering, the first recorded in Germany, followed by others at Augsburg and Munich.

PROVENANCE: Although it cannot be proved that drawing and print have always been associated with this set, it seems probable that they were. If so, the copy probably belonged to Joachim Ernst, Markgraf of Brandenburg-Ansbach (1583–1625), leader of the Protestant Union but a close friend of Bishop Johann Conrad (see above pp.6, 39–40). The accuracy and sobriety of the colouring accords with the horticultural interests and Protestant beliefs of the Markgraf. The binding is sober and somewhat old-fashioned. The book may well have remained in the family library until, with the merging of the principality of Ansbach with Prussia, the library was largely transferred to the library of the University of Erlangen in 1805. Three copies of the *Hortus Eystettensis* came from Ansbach, two dated 1613 and one 1713. Of these only the last now remains, and the two others are recorded as 'abgegeben am 1.7.1839'. Despite its colour, this may be one of these, since the University already had a coloured copy, once the famous possession of the University of Altdorf (see Er). If not 'given away', the copy may have entered the trade, either via C. H. Beck at Nordlingen or J. M. Heberle at Cologne; the University had dealings with both. This copy was certainly acquired by Maria Theresa Earle from Baer at Frankfurt and she records its purchase, with provokingly little detail, in *More pot-pourri from a Surrey Garden* (1889), p.92. It is probable that the repairs and re-endpapering were done at this time. It later passed to the De Belder collection, and was sold at Sotheby's, 27 April 1987, lot 23.

Er

LOCATION: Erlangen, Universitätsbibliothek, Trew B1[b].

BINDING: One volume, reddish-brown tanned hide, central circular 'sunburst' of multiple tools, panel formed by cornucopia roll within multiple fillets, joined at corners to outer border with pomegranate and scroll roll within similar fillets, all gilt. Spine with 6 raised bands, gilt oval ornament in centre of compartments, plain dark blue edges.

PRELIMS: Title, dedication IOHANNI CHRISTOPHERO only (blank paper pasted over final half page), portrait only (enclosed within thin pink line), French and Belgian/Dutch privileges.

SEASON TITLES: None.

CONTENTS: Plates only as rectos, no text, neatly foliated in contemporary manuscript, and with the German names written in more than one formal hand, not unlike that in BL, next to or above the engraved Latin names.

PAPER: Laid, 'grapes' watermark, 30mm between chain-lines, throughout.

COLOUR: Title background, greeny-blue slate background, trace of thin border of black at back edge, title cloth plain yellow, reverse bright green, detail black. The plates are unsigned, but conform to the consensus of early copies, without significant variation.

SPECIAL FEATURES: The German names, clearly written non-consecutively, vary from those given in the printed text.

PROVENANCE: This is of particular interest, since the copy has been in professional possession or use since, if not the first, at least very early ownership. It has recently been demonstrated that the first recorded owner was the Nürnberg apothecary and author, Johann Leonhard Stöberlein (1636–1696), who bequeathed his medical books to the University of Altdorf. There it was given pride of place in the library. According to Johann Martin Trechsel, *Amoenitates Altdorfinae* (1720), 'auf einer grünen Tafel an dem Pult liegt der schön kostbar elaborierte und mit sehr schönen Farben nach dem Leben auf das künstlichste illuminierte Hortus Eychstetensis [*sic*] in rotes Saffianleder mit Messing beschlagenen Buckeln gebunden'. This is clearly illustrated in plate 16 of the same work, the first to describe and depict any known copy of the work. The book is described again, in unmistakable terms, by Christoph Jakob Streck in 1750 and again by Christoph Gottlieb Murr in 1791. It may be that the possession of such a famous treasure led to the subsequent disposal of the Ansbach copies. At a later stage, for understandable reasons, the copy was aggregated with the all but comprehensive collection made by Christoph Jakob Trew, and given the shelf-mark which it now bears. Its earlier history, as noted, has only recently been recovered (see K. Wickert, *H.E.*, 123–5).

Eich

LOCATION: Eichstätt, Universitätsbibliothek.

BINDING: Three volumes, vellum over pasteboards, double panel of triple and double fillets enclosing large lozenge centrepiece, rebacked, with earlier double letterpieces (perhaps contemporary with binding) laid down on near spines, reading HORTUS/EYSTETTENSIS and VOL. I[II, III]/DEPICT., buff speckled edges, German, $17C^2$–$18C^1$. Bookplate of 'Earl of Aylesford, Packington, Warwicks', preserved and laid down on present endpapers, contemporary with rebacking, $19C^2$.

PRELIMS: Title, dedication IOHANNI CONRADO, 'Decanus . . . lectori', 'Ad nobilem . . .', blank/French, Dutch/Belgian privileges, portrait and arms.

SEASON TITLES: All four present.

CONTENTS: Plates alternately on versos and rectos, with the text backed up, facing the relevant plate, 'Verna' as volume', 'Aestiva' as volume II, 'Autumnalis' and 'Hyberna' together in volume III.

PAPER: Plain laid, ±32mm between chain-lines.

COLOUR: Title background dark brown, thick black border, title cloth chrome ellow, crimson reverses, detail left black. Portrait and arms transposed (Besler right and arms left) on a yellow background; arms correct. The colouring of the plates conforms with the 'early' consensus in general, but with some shared characteristics with later copies. Unique details are rare and generally insignificant, but note that 176 ii and iii are transposed in this copy only. No colourists' signatures.

SPECIAL FEATURES: The tulip plates are arranged in a very various order, in sequence: 67, 69, 71, 70, 73, 68, 72, 74, 75.

PROVENANCE: Presumably in Germany until *c*1700; then Heneage Finch, 4th Earl of Aylesford (1756–1812), member of the Society of Dilettanti and a notable artist and engraver who etched his own bookplate. He is known to have travelled in Italy, and may have acquired this copy *en route*. He was also a considerable book collector; the Packington Library was sold at Christie's, 6–16 March 1888, where this copy was lot 172 (Quaritch, £5 5s.). The need for a copy at Eichstätt had been emphasized by the publication of Schwertschlager's monograph, and was clearly known to Quaritch who sold this copy to the Episcopal Seminary at Eichstätt in 1892 for £9 10s; thus it, although sophisticated before or after Lord Aylesford's acquisition, represented an answer to a felt need. Transferred to the University Library in 1981.

M

LOCATION: Mainz, Stadtbibliothek, 640 gr.f.1.

BINDING: One volume, tawed hide over thick paste-board, eight blind roll-tooled panels enclosing small central void, plain spine, plain edges and endpapers.

PRELIMS: Title, two leaves torn out (the two dedications, with portraits, to Bishops Johann Conrad von Gemmingen and Marquard Schenk von Castell).

SEASON TITLES: Present (text plates only, without the border).

CONTENTS: Plates only, on rectos, without text, except the list of contents of each season placed after the title and before the text.

PAPER: Serpent, occasionally crown watermarks, 26–28mm chain-lines.

COLOUR: Title uncoloured. Plates systematically coloured, but only as far as plate 19, where the leaves and the flowers of 19 iv only have been finished. Although it is hard to judge from so small a sample, the colour variations are interesting, conforming in some instances (plates 5, 7, 15) to the 'later' copies, in others (13, 16) to the earlier. The leaves are all painted a pretty uniform faded green bice with little attempt at shading or variation of colour. Plates 4 and 10 have white roughly washed over the flowers.

SPECIAL FEATURES: From plate 362, German names have been added sporadically in a contemporary hand. There is a manuscript index at the end, inserted loose, but apparently there since made.

PROVENANCE: The index concludes 'Finitus hic Index â me Michaele Voss Med.D^{re} & professor 4° Septembris Ao. 1695', not, presumably, the first owner. 'E.B.V.M.' stamp on title, and 'iure emptionis me possidet carthusia moguntina 1715'.

BNP

LOCATION: Paris, Bibliothèque Nationale, Res.S.60.

BINDING: Two volumes, dark slate-blue morocco over thick paste-boards, arms of France in centre of boards, thick and thin fillet border, all gilt, elaborately gilt spine, gilt edges, marbled endpapers, $18C^2$.

PRELIMS: Title, dedication IOHANNI CUNRADO, 'Decanus . . . lectori', 'Ad nobilem . . .', 'Autores'/portrait and arms.

SEASON TITLES: All present.

CONTENTS: Plates on versos, printed text on rectos, backing throughout, indexes bound up after each part; these begin (unusually) with 'Hyberna'.

PAPER: Unwatermarked, 32mm chain-lines.

COLOUR: Title background, strong mid blue, title cloth white shaded crimson, title and D of DILIGENS painted gold, arms wrong (tinctures reversed); first two lines of dedication in gold, painted initial here and elsewhere in prelims and painted tailpieces to indexes; separate rectangular backgrounds to portrait and arms, bright crimson with alternating gold sprays of palm and laurel; the arms are wrong (the field is yellow, not blue). The lettering of the season titles is painted gold; the colouring is more or less uniform. The plates evince some significant changes from the earlier copies, but the overall difference is more marked: the colours are brighter, notably the blue, but there is a significant restriction of palette (the same pink, magenta, blue, vermilion, chrome, with green bice for the leaves, recur). The colours on Plate 90 iii and iv are transposed; this may be due to the fact that this copy was coloured from written instructions rather than an exemplar, or copied from another copy thus coloured (see S).

SPECIAL FEATURES: Ruled throughout with three lines, the inner blue, the outer two red with the space between painted yellow (gold in the prelims).

PROVENANCE: Inscribed 'CAROLUS LABBE, GABR. FIL. BITUR./1648' at foot of title. Charles Labbé de Montveron, the famous scholar and editor of the Pandects, admired by Joseph Justus Scaliger. Although he was born and died in Paris, he came of a family long established in Bourges (see Comte de Toulgoet-Treanna, 'Les recherches de noblesse en Berry', *Mémoires de la Société des Antiquaires du Centre*, XXIV (1901), 100. Bibliothèque Royale by 18C (old shelf mark on old front flyleaf S.241/C.1).

MHN

LOCATION: Paris, Museum Nationale d'Histoire Naturelle, 8568.

BINDING: One volume, pale calf over pasteboards, thick and thin filler border, oval arms of Fagon impaling Philippes de Barac'h in centre of boards, all gilt, gilt spine, marbled endpapers, ? edges; $17C^2/18C^1$.

PRELIMS: Title, dedication IOHANNI CUNRADO, 'Decanus . . . lectori', 'Ad nobilem . . .', portrait and arms.

SEASON TITLES: All four present.

CONTENTS: Printed text on rectos, plates on versos throughout, indexes?

PAPER: Unwatermarked, 32mm chain-lines, except portrait and arms leaf which is on paper with the 'grapes' watermark, 30mm chain-lines.

COLOUR: Title background sky blue, with white clouds, title cloth purple shaded dark, reverse similar, large letters painted gold, small silver, arms wrong (1 & 4 croziers or instead of argent, 2 & 3 vert & argent instead of azure and or); portrait and arms without background, legends to both painted gold (arms wrong). Head and tail pieces in prelims painted, but not initials. Season-titles painted in diverse colours, all with plain centres and gold lettering; shell surround green for 'Verna', blue for 'Aestiva', mauve for 'Autumnalis', orange for 'Hyberna'. Here, and in the plates generally, colour has been applied so heavily as to obscure most of the engraved detail (Plate 120 i is an obvious exception), with a range of colours so different from all other copies that points of connection are hard to find. The closest match is with BNP and S (cf Plates 39 v, 90, 107 i & ii, 181–2), but there are also exceptions (e.g. Plate 94); there can be no direct connection, and the common ancestor must be at more than one remove. But the most striking difference between MHN and all other copies & not the colour values, but its density. In no other copy is the base engraving so thoroughly occluded.

SPECIAL FEATURES: Plates 92 and 131, and 322 and 323, are transposed.

PROVENANCE: Guy-Crescent Fagon (1638–1718) Louis XIV's doctor; armorial stamp (Olivier 1912) of his arms impaling those of his wife N . . . Philippes de Barac'h. Thence to the Muséum de Histoire Naturelle.

MC

LOCATION: Dispersed.

BINDING: No longer extant, but described in the 1986 Monte Carlo sale catalogue (see below) as 'demi veau blond usagé avec manque, plats recouverts de parchemin'.

PRELIMS: 'Titre, portrait' alone are recorded in the catalogue.

SEASON TITLES: Probably none: 'ne portant qu'un seul frontispice', according to the catalogue.

CONTENTS: Plates only ('367 planches . . . sans les légendes au dos des planches'), although as usual the letterpress part-titles were present, but bound as versos (as DeB) with the plate on the recto. Foliated throughout in a contemporary French hand (see below).

PAPER: 'Grapes' watermark, ±30mm between chain-lines.

COLOURING: Despite damage ('66 planches sont ou déchirées ou attaquées par le colori' according to the description, and the purchaser estimates the number as higher), this was a copy of considerable quality. Frontispiece (description based on catalogue reproduction) background sky blue, with broad gold border, title cloth white shaded crimson, detail black; 'portrait' not seen. The thirty-eight plates seen (a further plate, 322, was reproduced in the catalogue) show that the quality of the colouring was considerable. The incidence of variants suggests that, with S, it forms a bridge between the earlier copies and the later, and may be an earlier version than the other two French copies (BNP and MHN), perhaps derived from a common archetype (see App. ?). The red 'Primula Veris Anglicana' (128 ii) and 'Papaver spinosum' (288 i) are cases in point. The flowers of the left-hand 'Momordica' in plate 322 iii is, surprisingly, coloured light blue (see OS). The diversely coloured peppers (327 and 329) are unique, while the flowers of the plants on plate 361, all properly white, are coloured yellow, pink and green, apparently haphazard, a trait partly shared with MHN, which shows two flowers of 361 i pink, and the green flowers of 'Helleborus niger flore viridi' (362 i) yellow.

SPECIAL FEATURES: It is clear from the foliation that this copy was bound in a most unusual order, apparently reflecting the seasonal order of the normal sequence of the *Hortus Eystettensis* without following it. This may in part represent the order in which some of the plants and trees grew in a contemporary French garden, since the same hand that supplied the foliation has added notes to some plates which evidently record flowering times. The folio numbers (with the plate numbers used here in parenthesis) and the notes are given below:

6 (128)		28[?] (227)		
42 (25)		292 (318)		
58 (69)		305 (255)		
59 (71)		313 (9)	'Aoust'	
68 (59)		315 (308)		
96 (94)		324 (246)		
115 (115)		331 (212)		
184 (196)	'este Juin'	334 (271)		
193 (13)	'Juin'	336 (213)		
200 (239)	'Aoust'	342 (248)		
202 (272)	'Juin sur la fin'	345 (127)		
219 (226)	'juin'	349 (241)		
221 (225)	'May & juin'	350 (224)		
226 (12)		353 (361)		
234 (203)	'Juillet'	356 (327)		
247 (288)		357 (329)		
258 (24)		359 (328)		
265 (232)	'juillet'	360 (326)		
277 (220)	'juillet sur la fin'	361 (330)		

The plate reproduced in the catalogue was trimmed and its original number is not visible. The dispersal of the plates and the destruction of the binding (which was clearly unusual — the vellum sides may reflect the difficulty of finding a full hide big enough for the book — and may have been early) make it unlikely that the full structure of the book, and the use for which it was intended, can be recovered. One small detail of interest survives. The now disbound leaves show on the inner (left) side the untrimmed deckle edge, punctured with the holes made by oversewing, usually 32 in number, allowing for paper defects and tears.

PROVENANCE: An original French owner, perhaps in the second quarter of the 17th century. Sold by Sotheby's Monaco S.A. at Monte Carlo, 20 October 1986, lot 413, and bought by Burgess-Browning Ltd, by whom it was resold shortly after to W. Graham Arader III; it has since been broken up, as noted above.

S

LOCATION: with Heribert Tenschert, Bibermühle, Switzerland.

BINDING: One volume, brown hide, double gilt arabesque roll border, each enclosed by multiple blind fillets, lozenge-shaped central panel of multiple blind fillets, enclosing armorial device on oval panel (a chevron gilt with an azured arabesque tool on a plain field), inside corners of fleur-de-lys (the lily flower made up, with curved carnation tools to left and right), diamond shaped formal floral stamps in centre at top and foot and twice on each side; ten raised bands, with six petalled flower in centre of each compartment, lily flowers to left and right, over multiple blind fillets. Plain edges and endpapers.

PRELIMS: Title, dedication IOHANNI CHRISTOPHERO, French and Belgian/Dutch privileges, portrait and arms.

SEASON TITLES: None.

CONTENTS: Plates on rectos throughout, except where backing part-titles, which appear as rectos with the following plate on the verso. No other text. Foliated throughout in a neat 19C[1] hand.

PAPER: 'Grapes' watermark, 28·5–30mm between chain-lines.

COLOUR: Title background sky blue (without border), titlecloth white (plain), shaded yellow-brown, arms wrong (tinctures reversed); no extra background to portrait and arms, the latter wrong (red angel, yellow field). The plates differ in a number of important respects from all other copies (e.g. Plate 4 i is, uniquely, red), but the most interesting feature is the number of instances in which the colour of adjacent flowers are transposed: 39 ii & iii, 72 i & v, 90 iii & iv, 121 ii & iii, 179 ii & iii, 196 ii & iii, 212 i & ii, 270 i & iii, 298 iv & v, 315 i & ii, 364 i & iii and 365 i & ii. One or two such instances (see App. E) might be accidental, but so many (including some directly contradicted by the captions) cannot be: it proves that S was coloured not from an exemplar, but a written colour chart, numbered as the captions are numbered. The colourist's mistake was to read the list from left to right, instead of following the more arbitrary, greater to lesser, sequence required by the plate. Several other features connect the colouring with that of MC, MHN and BNP, and in one instance to B2.

SPECIAL FEATURES: On the front free endpaper, oval arms (party per fess, azure and gules, a bar or) in a floral border on a cut out oval; to left and right, the large initials 'A' and 'V'. Plate 30 is placed where plate 24 should be; plate 41 is bound after 179 (q.v.); plates 50 & 51, 53 & 54, 55 & 56, 57 & 58, 59 & 60 and 72 & 73 are transposed; plates 80–85 are missing; plates 148 & 149 and 177 & 178 are transposed; plate 179 (with 41) is bound after 168. It is difficult not to believe that the number of adjacent plates transposed has something to do with the transpositions of colours in the plates; perhaps the list itself was defective in some way. A print entitled 'Vera, e natural effigie della pianta Indiana chiamata Maraco, Grandilla, e fior della Passione D.N.S.', with an inscription by Tobia Aldini, 'semplicista' to Cardinal Farnese, dated from Venice, 20 July 1620, is pasted to rear fly-leaf.

PROVENANCE: The arms (correctly painted; the chevron on the binding is probably due to a misunderstanding of the three-dimensional style of baroque heraldic depiction) are those of Vendramin, and the initials make it clear that the first owner was the great art collector, Andrea Vendramin, *fl*.1615–48. The catalogue of his pictures was published by Tancred Borenius, *The Picture Gallery of Andrea Vendramin* (London, 1923) from B.L. Sloane MS.4004. His complete set of catalogues was sold at Amsterdam in 1702 from the collection of Albert Bentes (*Bibliotheca Bentesiana*, Amsterdam, 1702, iii. 111, No.49); vol.XIII, 'de libris Chronologicorum universalium figuris & coloribus ornatis. De Iconibus aere et ligno incisis Alberti Aldegravii & aliorum Pictorum insignium. De Animantium, Piscium & Avium cujusvis generis formâ & Historiis. Plantarum et florum nobiliam Viridario . . .', was in the Zaluski Library, Warsaw (Jacopo Morelli, *Notizie d'opere di disegno* (Bassano, 1800), 237). Inscription listing missing plates (in English of a sort, but in a non-English hand), 19C[1]. Bought by Kurt Bösch at Ernst Hauswedell's auction rooms in Hamburg, sale 105 (3 June 1961), lot 87 (DM.23,500).

O

LOCATION: Österreichische Nationalbibliothek 68 A 24 (ES 388).

BINDING: Two volumes, red morocco gilt, German or Austrian, 17C[2] Broad gilt floral roll outer border within narrower rolls, enclosing central panel of elaborate floral gilt roll enclosed within two narrower rolls, separated by gilt fillets; larger narrow mandorla-shaped centre, made up of large floral tools (two tulips, carnation, poppy and iris) on cross-shaped stems, large cornerpieces incorporating urns and floral ornaments.

PRELIMS: Title, dedication JOHANNI CUNRADO, 'Decanus . . . lectori', French and Belgian/Dutch privileges, portrait and arms.

SEASON TITLES: All four present.

CONTENTS: Plates unbacked, alternately on versos and rectos, interleaved with text, backed up so as to face relevant plate, 'Verna' followed by 'Autumnalis' and 'Hyberna', 'Aestiva' in second volume, no indexes.

PAPER: Plain laid, 32mm between chain-lines.

COLOUR: Title background pale blue, wide orange border, gilt and shaded, title cloth crimson, shaded darker, green reverse, detail black, arms correct; general background to portrait and arms pale blue, orange border, arms correct, inscription round portrait crimson (not yellow), but otherwise close to early copies throughout. The portrait and arms are on a blue background with an orange border, like the title-page, and the arms are correct in every detail. Besler's name and date have been painted out on the titlepage, and replaced by 'Magdalena Fürstin Colorum Umbrarumque amictu alternavit. Noribergae', a curious phrase which seems to refer to the need to balance the colour applied with the already engraved detail, which may already have caused difficulties (see BSB). The colouring of the text is again very close to the early copies; there is one instance of transposition of colours (16 ii & iii), but otherwise variants are few. The plates are variously signed 'HSF' or 'HF' (plate 346 is signed in the latter form and dated 1671) and 'MF' (plate 359 in full 'Magdalena/Fürstin Illū/A° 1677'). The work of Hans Thomas Fischer (1603–1685) and his pupil Magdalena Fürstin (1652–1717) is described above (pp.45–46).

SPECIAL FEATURES: A label pasted to the front paste-down endpaper of volume 1 reads:

> Gegenwertiger Hortus Eystettensis, und schon illuminirung Jungfrau Magdalena Fürstin fünff Jahr gearbeitet hat, und der Römisch Kayerliche Majestät Hofbibliothek in Wien allunterthangst angeboten umb vier hündert Reichsthaler. A. 1678 d. 29 Augusti auch aussbezahlt worden.

The same hand has noted the number of plates in each volume on a smaller label.

PROVENANCE: Bought by the Imperial Librarian, Peter Lambecius, on 29 August 1678 for the Imperial Library at Vienna for 400 Reichsthalers, on the evidence of the label quoted above. The book was famous and was described by Johann Gabriel Doppelmayr in his entry for Magdalena Fürstin in *Historische Nachricht von den Nürnbergischen Mathematicis und Künstlern* (Nürnberg, 1730), p.270, which records the part of Lambecius in the transaction. The inscription may be in his hand. The book has been in Vienna since it was acquired by the Hofbibliothek, now Nationalbibliothek.

BSB

LOCATION: Munich, Bayerische Staatsbibliothek, Rar. 1093 (formerly 2° Phyt. 18).

BINDING: One volume, old reddish brown hide over old boards (pastings of pulpboards, apparently replacing older wooden boards, since the bevel still remains), oblong arabesque centrepiece, enclosed within triple border of rolls enclosed within fillets, with inside corner-pieces, all gilt. Traces of two pairs of clasps on foredge. The binding resembles those of N, W and Er, and cannot be later than mid 17C; roughly rebacked, probably 19C.

PRELIMS: Title, 'Ad nobilem . . .', dedication to IOHANNI CUNRADO, 'Decanus . . . lectori', Belgian/Dutch and French privileges, 'Autores'/portrait and arms.

SEASON TITLES: 'Verna' only.

PAPER: No watermark, 32mm chain-lines.

CONTENTS: Letterpress text on rectos, plates on versos throughout to plate 99.

COLOUR: Frontispiece background uncoloured title cloth uncoloured, arms wrong or perhaps incomplete (2 and 3 barry azure and argent, instead of or); portrait has green background, arms red, the latter incomplete at the back margin (therefore painted after the book was bound). The arms are again incomplete. The one season-title is fully but hastily coloured, except for the centre panel. As in O, the colour scheme follows the 'early' copies closely; it is surprising to find plate 31 i white (uniquely), since the plate is signed. Plates 30 and 92 are signed 'MF', 31, 67, 69, 70, 77, 80, 81, 82, 87 and 88 'HSF', i.e., Magdalena Fürstin and Hans Thomas Fischer (see O).

SPECIAL FEATURES: Plate 5 has been torn out, plates 6 & 7 transposed, and 29, 32, 67, 69, 70, 77, 80, 81, 82 and 87 missing. The rough and partially incomplete colouring of a copy not meant for colouring, in already bound state, is puzzling. It may represent a trial by Fischer and Fürstin for O, or more likely an unsuccessful attempt to repeat that success. It was the only coloured copy seen by Schwertschlager, although he had heard of the copy formerly at Altdorf (Er), and he describes it in some detail.

PROVENANCE: Bound, probably before 1650; coloured by Fischer and Fürstin, presumably about contemporaneously with O, i.e. c1670–80; old stamp of Bibliotheca Regia Monacensis on verso of title, and thus now in the Bayerische Staatsbibliothek.

LU

LOCATION: Leiden, University Library, Plano 50.E.1.

BINDING: One volume, half calf with flat spine (title gilt), fine brown speckled edges, plain endpapers.

PRELIMS: Title, portrait and arms (laid down), dedication IOHANNI CUNRADO, 'Decanus . . . lectori', 'Ad nobilem . . .', French and Belgium/Dutch privileges.

SEASON TITLES: All four present.

CONTENTS: Printed text on rectos, plates on versos throughout, no indexes.

PAPER: No watermark, 32mm chain-lines.

COLOUR: Title background plain, title cloth plain, shaded ochre, arms plain; portrait and arms plain. The 'Classis Verna' title is incompletely coloured, but with great skill and subtlety, the right-hand pillar and pedestal being marbled in contrasting patterns, the panels on the pedestal gilt; a black and white marbled floor is painted in the foreground and the pyramid of bottles individually high-lighted to simulate glass. The plates are all rather grimy and the colouring thin (gouache colours applied so diluted as to become water-colour); this is to some extent a necessity, to counterbalance the dark details on the plates. In general, the colouring conforms to the 'later' pattern, although with some interesting exceptions, e.g. at 13 i and 75 ii, which both reflect the older tradition. On the other hand, no attempt has been made to render the 'Squammata' (Plate 15), the blue gentian (113 iii) has been coloured yellow by confusion with the neighbouring buttercup, and the curious transposal of the inner and outer petals of the poppies at 292 ii & iii indicates misunderstanding of the exemplar and startling ignorance of nature. Many of the leaves are rendered in the same etiolated yellowish green as in Er2, S, MHN, Ellw, and (to a lesser extent) BNP and MC. All the cactus leaves (plate 354ff) are unshaded in plain pale colours. There is some show-through, e.g. on plate 345.

SPECIAL FEATURES: The sections on spring and summer trees are bound after the main 'Verna' and 'Aestiva' sections, not before as usual. Plate 182 (the top section of plate 181) is mis-bound between plates 88 and 89, and an extra print of 182, uncoloured, appears in the correct place.

PROVENANCE: Before the frontispiece, a handsome calligraphic leaf records the bequest of the book: 'Dit Boek genaamd de HORTUS EYSTETTENSIS is gelegateerd geweest by den Wel Edelen Heer HIERONYMUS VAN BEVERNINGK, Heere van Teylingen etc etc etc aande Bibliotheek van de Universiteit tot Leyden in erkentenis vande eere die syn wel Edeleheeft gehad, veele iaaren tebekleeden de CURATEUR-PLAATSE vande selve Universiteit. 1690.' In the University Library since.

BR

LOCATION: Brussels, Bibliothèque Royale Albert I, VH 6227 E (LP).

BINDING: One volume, vellum over pasteboard, boards with double broad arabesque borders with flowers at corners, large inner corner-pieces and large lozenge-shaped central ornament, all gilt; flat spine gilt in compartments formed by roll-tool, flowers at centres with four floral arabesque tools towards corners, plain edges and endpapers.

PRELIMS: Title, 'Verna' season-title, portrait and arms, in that order.

SEASON TITLES: All present, but only 'Verna' and 'Aestiva' coloured.

CONTENTS: Plates only on rectos throughout no printed text except part-titles (on versos); foliated at foot and the plants numbered in 19C pencil.

PAPER: No watermark, 32mm chain-lines.

COLOUR: Title only partly coloured and much faded: background and title cloth uncoloured, arms wrong (1 & 4 crozier or not argent, 2 & 3 tinctures reversed); the only parts to be fully coloured are the swags of fruit to left and right of arch; portrait and arms have no background. Season-titles are both very palely coloured. The plates are generally thinly coloured, with some details left uncoloured. This is in part a necessity due to the blackened state of many plates, particularly plates 67–77. On the other hand the pinks (plates 308–9) are coloured with considerable care. The flower of 30 iii has been erroneously coloured yellow. The number of transpositions (second only to S) suggests that the colourist had an imperfect exemplar, supplemented by written instructions.

SPECIAL FEATURES: Two separate paper stocks are noticeable, one thick, the other thin; prints on the former are generally cleaner, in particular that used for part-titles; the latter was difficult to colour (e.g. plate 36, where colour has permeated). Plate 68 has been misbound after 70, and 148 is missing. There are streak marks due to imperfect wiping of the plate on plates 39 and 153. The part-titles which should have been printed on the versos of plates 319 and 361 have been lettered in manuscript. All this suggests a set largely made up of sheets discarded after the plates were printed and made up later.

PROVENANCE: Armorial bookplate (arms: a chevron between three trefoils, on a chief an eight-pointed star, crest: a helmet surmounted by a swan) of Hinlopen family of Hoorn and Amsterdam; later bookplates (on front paste-down endpaper and portrait leaf) of Charles van Hulthem (1764–1832), with a long inscription in his hand on the former. Bought with his collection by the Bibliothèque Royale in 1836.

Ellw

LOCATION: Ellwangen, Ellwangen Geschichts- und Altertums verein, on exhibition at the Schlossmuseum, Ellwangen.

BINDING: Two volumes (of three), tawed hide over thick pasteboards, blind-tooled, triple roll border on boards, each roll enclosed by triple fillet, double roll inner panel, similarly enclosed, corners joined to inner corners of border, large lozenge centre piece of flower, scroll and wreath tools, guide cross marks made with a stylus in centre, eight raised bands on spine, marks of four clasps (now missing) at top, tail and two at foredges, plain edges and endpapers, $17C^2$.

PRELIMS: The first volume is missing.

SEASON TITLES: 'Aestiva', 'Autumnalis' and 'Hyberna', uncoloured.

CONTENTS: Text on rectos, plates on versos throughout.

PAPER: No watermark, chainlines 32mm.

COLOUR: Colour has been applied to plates 140, 141, 144, 149, 151, 165, 166, 169, 170, 185, 322, 335, 337 and 367. It is only partial, sometimes only a single flower or leaf, but it has been carefully done, and clearly from a model. Where there are variants, the colouring here reflects the 'early' version, not the later. This is particularly notable in plates 144 and 149. The four flowers painted on 'Iasminum indicum' (plate 335) correspond exactly with those in the 'early' copies. The pale yellowish green leaf of the 'Momordica' (plate 322 iii) echoes the same sort of colouring in Erz, which it resembles still further in the 'part for whole' colouring.

SPECIAL FEATURES: Besides lacking the first volume, this copy wants plates 189–252, all roughly cut out from volume 2. It is thought that this vandalism took place at the end of World War II, when the book was removed from its war-time hiding-place. There is a light but talented drawing (c1800?) of 'Convolvulus tricolor' pasted to the back of plate 157.

PROVENANCE: Bequeathed to the Ellwangen Geschichts- und Altertumsverein by Franz Joseph Rathgeb (1835–1910), apothecary of the town, who probably inherited it from his father Johann Baptist Rathgeb (1796–1875). There is an early 19C shelf-mark '981 C' in both volumes, which suggests that any further connection with earlier Ellwangen natural historians, such as Joseph Alois von Fröhlich (1766–1841), is unlikely. On this, see Otto Häcker, 'Eichstätt und Ellwangen als Pflegestätten der Pflanzenkunde', *Ellwangener Jahrbuch* 1926–8, 111–125; it may be inferred from the fact that Häcker reproduces the title-page of the *Hortus Eystettensis* from a previous reproduction in the Mittenwald *Illustrierter Apothekerkalender* that volume 1 was already missing when he wrote. See Karl Fik, 'Der sogenannte "Hortus Eystettensis"', *Ellwangener Jahrbuch*, X, 310–11.

OS

LOCATION: Mellon Collection, Oak Spring, Virginia.

BINDING: One volume, half calf, speckled paper sides, flat spine, red lettering piece, plain edges and endpapers, German, $18C^2$.

PRELIMS: Title dated 1713, preface by Johann Georg Starckmann, headed 'Spectatori et Lectori'.

SEASON TITLES: None.

CONTENTS: Text on rectos, plates on versos, with alphabetical list of contents for each season.

PAPER: Poor white laid, without visible watermark, chain-lines c 30mm.

COLOUR: Title background pale blue-white without border, title cloth white shaded crimson, reverse bright yellow, arms of Bishop Johann Anton I von Katzenellenbogen. The plates are coloured with great skill, watercolouring quickly applied to prevent the risk of spreading due to the poor quality of the paper. The colouring in general conforms to the 'later' pattern, sufficiently closely to suggest that some model was available to the colourist. Plates 13 and 15, for example, relate to MHN, S, and MC, and to M and S, respectively (see App. B). Others, however, conform to the early pattern, notably the 'Ficus Indica' (plate 359). Besides these, there are some apparently unique features, such as the completely variegated 'Iasminum Indicum' (plate 335); the rose-red colour of 175 iii 'Consolida regalis multiplicata violaceo colore' may be an error, but the orange Iris at 197 iii is too carefully drawn for that to be the case, and the detail of the great sunflower (Plate 204) is too complex to be a copy; it must stem from a new drawing from nature. The copy shares one curious error with MC: the flowers of the right hand 'Momordica' (322 ii) are coloured blue (in MC, the other, 322 iii, is similarly mis-coloured).

SPECIAL FEATURES: Slight waterstaining in the head-margin has been carefully removed. There is a smudge of green colour on the recto of plate 226 (part-title for 'Aestiva' VII).

PROVENANCE: Sold at Sotheby's 15 April 1988, lot 42. Bought by Heywood Hill for the Oak Spring Garden Library.

Sp

LOCATION: Speyer, Pfalzische Landesbibliothek, 1b 188 rara.

BINDING: One volume, recent half-calf with new endpapers; evidently a recasing, since the original 18C blue sprinkled edges have been preserved.

PRELIMS: None.

SEASON TITLES: None.

CONTENTS: Text ('1713' edition) on rectos, plates on versos, no contents lists.

PAPER: Poor white laid, chain-lines about 30mm apart, no other watermark visible.

COLOUR: Plates 1, 111, 113, 114 (ii only), 115, 116, 119 (iii only), 125, 127, 128 (ii only), 129 (i & ii only), 130, 131, 132 (ii & iii only) and 134 have been coloured, sometimes only in part. Except for the fact that this small selection is almost a sequence, it would be possible to believe that it was coloured from nature without reference to any model. In that case, however, one would expect a more random selection. Whatever the source, the treatment is quite careful: the colouring of the reed *Acorus calamus* (Plate 125 i) uses several shades of green, and the treatment of the Asparagus (Plate 134 i) is unique.

SPECIAL FEATURES: Plate 141 missing; the difficulty of painting on this paper is evinced by show-through as plates 127–8.

PROVENANCE: Bought by the Pfalzische Landesbibliothek in 1956 with the library of Schuldirektor Emil Haas, a total of 2170 volumes for DM.2500; Haas had inherited it from his uncle, Georg Hook, who died in 1939.

APPENDIX A

Documents

1. Philipp Hainhofer's account of his journey to Eichstätt[1]

A Short and Trustworthy Account of how I, Philipp Hainhofer,
Citizen of Augsburg, carried out my Journey to Eichstätt
and Munich, and how the Correspondence between
the Royal Houses of Bavaria and Pomerania started.

After Sunday in mid-Lent [13th March] in the year 1611, His Highness, Lord Wilhelm, Count Palatine of the Rhine etc. journeyed to Augsburg and put up at the Monastery of the Holy Cross, residing there until several days after Easter. While here he sent for me on several occasions and graciously conversed with me about art and other subjects. My words turned, amongst other things, to His Serene Highness, Lord Philipp II, Duke of Stettin, Pomerania, of the Kashubes and the Wends etc.; and I most humbly mentioned that His Serene Highness was not only a learned and highly judicious prince, but also a lover of virtue and the arts who took pleasure in corresponding with learned and virtuous men of high or low estate. He read and wrote on many topics himself, was unflagging in his diligence, and to be compared to His Highness in many laudable ways. Thereupon His Highness inquired of me whether I was well-acquainted with His Serene Highness, how old he was, what faith he belonged to, what he looked like and what opportunities I had to send him the letters and other items. Of all this I gave a fitting account, and let His Highness see His Serene Highness's portrait in gold and amber as well as other fine gold coins and rare objects His Serene Highness had been kind enough to send me. His Highness [p.16] replied that the portrait showed a pious prince who, with his long Nazarene hair, would have looked exactly like an image of Christ the Saviour had he not had the sides of his beard trimmed in the German fashion. He continued: 'Do you think that His Serene Highness would be pleased to correspond with me, since I am no longer a ruling prince and, so to speak, lead a life secluded and remote from worldly affairs? With what might I initiate the exchange of letters and send His Serene Highness a worthy greeting?' To this I replied: 'My most humble opinion is that he would welcome the opportunity. For Your Highness has learnt much by travelling through numerous countries, has seen many peoples, and made yourself known far and wide. Since you exchange letters with kings and princes it does not behove me to order Your Highness to send a gift. However, I would humbly hope that if the presents Your Highness sent His Serene Highness included copies of the pictures of birds, fish and animals which you had collected and sketched some years ago these would be very well-received, as would plans of your various summer-houses. For His Serene Highness graciously wrote to me some time ago, asking me to obtain copies of Your Highness's and the Bishop of Eichstätt's drawings of animals and plants, since he wishes to have the same copied all and brought together in one book. Further he wrote that he was in the process of building a summerhouse, for which the foundations were already being laid, and requested me to take great pains to send the designs and plans of the Duke of Württemberg's summer-house in Stuttgart as soon as possible. Nor did he think for one moment that the Duke of Württemberg, when told they were desired for the Duke of Pomerania, [p.17] would refuse this request, even if His Serne Highness did not consider so slight a matter worth the trouble of writing to the Duke of Württemberg about personally, as he would not deny Württemberg anything he desired from Pomerania. On my most humble request permission to copy the designs was obtained from the Duke of Württemberg through the good offices of Dr. Hieronymus Bechler, J. U. D. and Royal Württemberg Privy Counsellor, my dear, trusted friend and for four years my teacher. I have been told that work on the same is well under way.

Your Highness would also make the Duke of Pomerania very happy if you now sent the plans of your summer-palaces, as also the ponds, Belvedere, the grottos, the antique statuary, various halls and other buildings, so that His Serene Highness in Pomerania could take details from each and incorporate them into his intended building, especially as he wants to erect an Art Cabinet next to it.' Whereupon His Highness replied that he would have work started on the animal sketches and that, should I be unable to, he would persuade the Bishop of Eichstätt to release his. He only demurred on the subject of the building plans, since neither he nor the lord his son ever allowed them to be disclosed or revealed to another potentate.

However, His Highness indicated that if I thought a special service would thereby be rendered this good prince, he would consider the matter further and see whether some piece of work might be produced *en relief*, since not everything could be copied flat with enough skill to convey an adequate impression of it. Furthermore, as he wished to be able to render this prince various services, the latter being such a great lover of art and virtue, he intended, in God's name, to begin the correspondence through me, [p.18] since I enjoyed the trust of this and other princes. Moreover, he intended to continue it through me, possibly sending a letter himself should a suitable occasion arise and referring to me with the gifts he sent. However, he said he would relate all this to me at greater length on his return to the court in Munich; in the meantime I should merely await developments.

Our conversations about the Duke of Pomerania lasted up to four hours on three separate occasions; and His Highness was particularly pleased by the fact that His Serene Highness had the year of his [Wilhelm's] birth in his name: PhILIppVs DVX PoMeranIae, that is, 1573, in which year he was born, which is singularly odd.

Moreover, I showed His highness a *Cornu Alcis* [staghorn] which the Electoral Prince of Brandenburg had prepared with his own hand without using fire, as well as a fine enamel crucifix from Limoges, which had been amongst the items sent to me by His Serene Highness. Because these objects appealed to him, I immediately and most humbly presented them to him, expecting others from His Serene Highness to replace them.

As soon as His Highness had arrived home he wrote to me — sending the letter via a personal messenger — about various *objets d'art* which had been turned on a lathe and about an ingenious little lathe that could be carried from room to room, screwed on to every table and driven by foot or by wheel. He sent three such letters and in the last, dated Schleissheim, 4th May, announced the following:

'Dear Hainhofer,
With my gracious greeting I send you in confidence what the Bishop of Eichstätt wrote and communicated to me. For I doubt that you will find or receive in those quarters what we did not begrudge the Duke of Pomerania, and yet the Bishop claims [p.19] to raise no objections to our sending someone to Eichstätt. Thus I ask you to consider at your leisure whether you would like to take the trouble of going there yourself and seeing how the land lies. If you did, I would give you, *tanquam subdelegato*, a letter of credence or missive addressed to His Lord the Bishop so that you may send further report to me hereafter and we can decide what more is to be done. I shall also await a drawing and description of the turned ivory piece which you sent the Duke

---

1. The text is here translated from that printed by Christian Haütle, 'Die Reisen des Augsburger Philipp Hainhofer in Eichstätt, München und Regensburg in den Jahren 1611, 1612 und 1613', *Zeitschrift des Historischen Vereins für Schwaben und Neuburg* VIII (1881), 1–360. Only the part relating to Eichstätt is given, with the relevant page numbers from Haütle's text; passages of lesser interest have been abbreviated.

of Pomerania, and request that nothing be done in vain or sent twice. I did not want to conceal the above matters from you, to whom I am so favourably disposed.'

Here follows a copy of the above-mentioned letter from the Bishop:
'Highborn Prince, accept my willing and neighbourly service. Gracious My Lord.

As to what Your Highness wrote to me concerning various sketches of diverse animals, plants, herbs or other curiosities and works of art, I may, with the best will in the world, not conceal that I in fact have no such pictures in my possession at present, especially of fish, nor do I have at the moment any four-legged animals, either alive or on paper, except for a few, albeit common birds which for the most part come from Munich. As for various flowers and garden plants, it is possibly no bad thing that some while ago I ordered sketches to be made of what had been observed in my own modest, narrow little garden. However, just now I do not have the drawings to hand but have sent them to Nuremberg, where they are to be engraved in copper and perhaps eventually published in the shape and form Your Highness can see from the enclosed.

[p.20] However, as to where the *rara* and curiosities mentioned in your letter can be acquired, I can tell you nothing more than that the garden plants were brought back, through the offices of local merchants, above all from the Netherlands, for example, from Antwerp, Brussels, Amsterdam and other places, and were then brought here. Should it please Your Highness to send someone to our court, I have not the slightest objection but would be glad to have him given an extensive tour of whatever we have here, for I am greatly desirous and concerned to show, to my utmost ability, His Highness and his estimable line every co-operation and service that is pleasant and dear to him, not only in this but in much more. Dated Eichstätt, 1st May 1611.

Your Highness's Most humble servant,
Johann Conrad'.

For which most gracious request by His Highness I offered my most humble thanks and showed my readiness to accede to his will in this matter. Thereupon His Highness, on Ascension Day [12th May], wrote, amongst other things, this despatch:

'Lastly you will find enclosed the letter of credence together with a copy of the same addressed to the Lord Bishop of Eichstätt, and along with it a note book so that, if it pleases you, you may go forth with it in God's name and render me account of what you find, thus enabling me to make the appropriate provisions. I remain graciously inclined towards you.'

The copy of the letter of credence to the Lord Bishop of Eichstätt written in His Highness's own hand, as well as the notebook, was worded as follows:

'Honourable, most dear friend in God, from Your Grace's kind reply I understand this much: that as far as the works of art are concerned you will gladly suffer me to send someone to Your Grace to whom you will then gladly show all you have in the say of such objects.'

Having received such an obliging offer, I am sending this envoy Philipp Hainhofer [p.21] to Your Grace as one trusted by to us in such matters, as a lover and connoisseur of art. From him Your Grace may learn more about what I meant in my recent letter with regard to various drawings. When a favourable opportunity arises may you grant him an audience, listen to him, and declare yourself in agreement with my trust in him and his report on various matters. I remain devoted and well-disposed towards you in friendship and all good things. Dated Schleissheim, Ascension Day 1611.

Your Grace's ever-devoted
Wilhelm.'

The note written in His Highness's hand was as follows:
'Record of what Philipp Hainhofer in my name and empowered by this letter of credence requests from the Lord Bishop and how he should conduct himself.

First of all, he should travel at his own discretion, however without incurring any costs greater than those which it is normal and reasonable to reimburse. As soon as he arrives in Eichstätt he is to send our letter of credence to the Lord Bishop and let him know the inn where he is to be found. When so requested he should present himself to out Lord Bishop; however he should await the latter's command and convenience at the inn.

When he is granted audience, he is to convey our friendship and neighbourly greeting to the Bishop and wish him health and prosperity. Over and above that he is to clarify our recent request to the Bishop, and ask him to respond to the questions we touched upon in our recent letter. He is to reiterate that in accordance with the Bishop's kind offer we have sent him, Hainhofer, to the Bishop to learn from His Grace how far he is willing to let us have what we have asked for. However, because our Lord Bishop may well wish to hear again at once what we are actually seeking in this matter, Hainhofer will, according to what he knows himself, be able to impart the necessary details and state what will be acceptable to my dear Lord of Pomerania. However we shall act as if Hainhofer were acquiring everything for us and leave the Duke of Pomerania out of the picture, especially since it may be more effective to do it this way rather than any other as we desire the drawings in our own name. Whatever is shown to you, Hainhofer, as a result you must memorise diligently and [p.22] when something fits in with your brief you should hint as much to the Bishop and desire that the same item be sketched for us.

However, should the Bishop make the excuse that he lacks people to do this for you, Hainhofer, must reach an agreement with the Bishop about what he is prepared to relinquish, so that we can appoint somebody to carry out that task and so that we may come to an arrangement with you ourselves for the copying to be completed at the lowest possible cost. However, if the Bishop has people, His Grace should acquiesce, since he may have reservations about sending the drawings out of house if they can otherwise be done at our expense. Thus we charge you, Hainhofer, to use your skill to arrange matters as the opportunity arises. Should it be completed, you could also request one, two or three copies of the Bishop's book of engravings which he mentioned in his letter. And should you, Hainhofer, succeed in bringing matters to a fruitful conclusion, you are to thank the Bishop on our behalf and inform him that we desire to respond in kind. Then you should pay your respects and, when the appropriate occasion presents itself, take your leave and make your way back to Munich in case we are to be met with there. You are to have yourself shown everything that is worth seeing in Eichstätt as well as at Schleissheim in case we ourselves have not yet arrived, just as you should announce yourself in our residence to our chaplain and secretary Herr Georg Schön, who will tell you all you need to know. Done at Schleissheim, 13th May 1611.

Wilhelm.

Yet again I thanked His Highness through the personal messenger he had sent for the gracious commission and trust bestowed in me, informing him that I wished to set off any day now and, God willing, report on my success personally in Munich.

Thus at six o'clock in the morning on 16th May in the year of Our Lord 1611 I rode out from Augsburg with a servant by the name of Hans Wachter, was in the suburbs of Bettmess by mid-day (which Bettmess is a market town divided into four parts and belonging to the four brothers Gumpenberg; the oldest brother must use the castle on the mountain, which is their ancestral home, as his permanent residence). In the evening I took lodgings in Neuburg, also in the suburbs, in the Golden Goose inn.

[p.23] Neuburg is a small town belonging to his Lord the Count Palatinate, where His Royal Highness the Count Palatine Philipp Ludwig holds court, which Royal Highness had a new church built in front of the palace). Item, handsome, very large and extensive earthworks around the entire town to enlarge and fortify it. However, many thousands of men would be needed to man and guard such fortifications in case of attack by the enemy. Court life is very frugal there.

Then on the 17th May I started on my way at four o'clock in the morning and reached Eichstätt at eight [am]. The town lies in a beautiful, deep valley; the castle, however, half-an-hour away on the top of a hill. In town I took lodgings at the Grape Inn on the market square and immediately sent a servant with His Highness's letter of credence to the palace, to Adam von Werdenstein, Councillor and Chamberlain to the Bishop. Through the servant I was informed that a messenger would soon be sent down to me, for whom I waited. After some time Wolfgang Agricola, Palace Steward, came with two servants in order to welcome me in His Lord Bishop's name. He summoned and accompanied me to court, where I was given the Prince's room looking out onto the moat (which was full of rabbit warrens). A guard was ordered for the room, who was to stand in waiting outside it day and night, as well as a courier to send hither and yon.

[p.24] Now when Agricola had conversed with me for a while in the room and had had my horse and luggage fetched from the inn, he left me for a time and a servant assisted me with undressing. Half an hour later Adam von Werdenstein came and received me once again in the name of His Royal Grace the Bishop, whose apologies he offered for not granting me audience straightaway and dining with me, however, he was indisposed. When the Bishop felt better, however, he wanted to speak to me that afternoon in the garden, where he intended to have himself carried. Consequently he had assigned von Werdenstein and Agricola to me and told them to be at my disposal the whole time, which order they carried out diligently. They dined with me downstairs, and in my room I was handsomely served on a silver service with many different dishes; confectionery; Italian, Spanish, Canary and German wines; and liqueurs. Before the meal I was entertained with great ceremony.

After the meal I was left alone for an hour, whereupon Agricola came and conversed with me for a while, and after him von Werdenstein. He greeted me in the Bishop's name and proffered his apologies since, contrary to his original intention, a congestion of the lungs had prevented him from being carried into the gardens in such windy weather. Instead he was forced to remain in bed in a warm room and wanted to see that he granted me audience early the following morning.

In the meantime he, von Werdenstein, and Agricola were to attend to my needs with the servants and the palace guard and give me a tour of the gardens, as then happened. We visited what must have been eight gardens situated round the palace, which lies on top of a hill and bears a fair resemblance to this engraved picture. Each of the eight gardens contained flowers from a different country; they varied in the beds and flowers, especially in the beautiful roses, lilies, tulips, among all other flowers. . . . The gardens were partly adorned with painted halls and pleasure rooms, in one of which halls stood a round ebony table whose surface and feet were inlaid with flowers and insects engraved in silver.

From the lower castle garden we went through the masons' yard and smithy to the quarry, where we saw the cliff on which the palace stands being blasted with gunpowder and large blocks extracted which were prepared for building, as approximately 200 men from Graubünden and Italians were constantly at work on it. Twelve heavy horses drag the stone uphill.

His Grace wants to turn the whole palace round and have it built from blocks of rock on top of the cliff. It is intended to roof over one side this summer, all of which will be covered in copper; and all together it will cost more than 100,000 florins. The gardens will then be turned round as well and levelled with each other all round the palace on the slope. On the side facing east an exquisite chapel is to be built, with all the windows nine feet high, without any panelling, not even timber-work, instead only a cornice from which to hang tapestries.

[p.26] And by the quarry a stream flows out of the rock which has been led round the whole of the palace mountain; it is called the Altmühl and yields excellent trout, pike, bullheads and even nice big crabs.

In the rock of the cliff can be found [fossilized] fish, leaves, birds, flowers and many strange things which Nature makes visible there.

Subsequently we went into the pheasant gardens, of which there are four different kinds: in the one are white pheasants, in the second speckled, in the third and fourth red ones, likewise cranes and other birds.

Just as the gardens are different, so do they also have different gardeners, as none infringes on the other's domain.

When we left the gardens I was once more accompanied to my lodgings and left alone inside for a while, then dined again; the Bishop greeted me through a gentleman-in-waiting and wished me a good night. The silver waiter, who came and served at table every time, cleared everything away again with his servants.

On the 18th May His Royal Grace again sent Agricola to me at half past six in the morning to wish me a good morning and ask what I desired to eat and drink for breakfast.

At seven o'clock, because I wanted no breakfast, von Werdenstein also came in His Grace's name and wished me a good day. He asked how I had slept and whether it would please me to have an audience then; if so, he and other courtiers would attend me. Thus I went with them to His Princely Grace's chamber, kissed his hand and while he still stood I said something along the following lines:

That His Grace would still recall what His Highness, Lord Wilhelm, Count Palatine etc., my most gracious prince and lord, had written about drawings of various birds, flowers and other *exotica*, and what answer His Grace had given His Highness: namely, that he would gladly suffer [p.27] someone to be ordered to Eichstätt when it pleased His Highness and that the same person would be presented on his arrival and a communication sent to His Highness. Since His Highness [Wilhelm] had gladly and gratefully accepted His Grace's obliging offer, he had sent me with the letter of credence already most humbly conveyed to His Grace and had graciously commanded me to convey His friendship, neighbourly greeting and best wishes for His Grace's health and well-being. He also further requested an opportunity for speech so that I could reveal more of His Highness's intent to His Grace, and desired me to wait and discover what His Grace would be pleased to have me shown. To all this His most esteemed Royal Highness should not fail to reply at the desired opportunity and with friendship and good will.

Thereupon His Grace gave an answer couched in the following terms: 'I rejoice at the gentleman's presence and dutifully thank His Most Serene Highness Duke Wilhelm of Bavaria, my gracious Lord, for his proffered greeting and the kind confidence he places in me. I wish I had that which His Royal Grace would seek from me. However, since I have nothing His Grace does not already possess, indeed, in more beautiful and better form, and since in addition the flowers, my most precious drawings, are now in Nuremberg (whither I was asked to send them by an apothecary who helped me lay out my garden and increase the number of flowers; he wishes to have them engraved in copper, printed, dedicated to me and to seek his fame and profit with the book), my Lord must suppose that he has caused this trip to be undertaken in vain. However I shall have everything faithfully shown, what little remains of it. Beforehand let us converse a while with each other.'

Then His Grace sat down, pulled a cover over himself and said he could stand no more as his feet were quite unwilling to support him any longer. Thus I had to sit down by His Princely Grace, cover myself as well and remain alone with him for half an hour.

The conversation dwelt on His Highness in Bavaria, his condition and life, on pictures and especially flowers [p.28], since His Grace told me that Beseler, the apothecary in Nuremberg, was at that time fully engaged in working on the book; that His Grace would publish it and had one or two boxes full of fresh flowers sent there every week to be sketched; how he always had tulips in five hundred colours, almost all different; and that this book would cost around 3,000 florins.

Since he did not have enough strength in his hands to open it himself, His Grace had me open a particularly fine writing desk which stood on a small table in his chamber, on the right-hand side as he sat at table. . . .

Then His Grace rang a small bell. Von Werdenstein and other

servants came in and were ordered by the Bishop to conduct me to the balconies in front of his room, to the dressing room and to the treasure chamber. On the galleries, in front of the lovely, bright windows with large crystal lights (so clear that the light shines into the room as if there were no panes and one is tempted to poke one's head out through them), stood various plants: red, yellow, brown and speckled pansies in flowerpots; and in the middle of the gallery violets, apricots, pomegranates, lemons, Amaranthus tricolor etc. in tubs.

On this balcony stand six large blocks of the kind butchers use for cutting. In them are placed dead trees and as His Grace was looking out of the window talking to me, I asked him what their purpose was. He answered me that in winter he had a flock of birds in front of his room, since he always had bird food scattered outside and the birds came in masses, sometimes two hundred at one time. They looked for food and sang together. He let them fly free, because if he caught them he would only drive them away and be robbed of his enjoyment.

A pipe, connected to a tank, was laid through the room and out of the window to irrigate the land beneath. All day long the water runs through the Bishop's own room as if it were a conduit and is conducted up hill through pipes.

Hainhofer was then given a conducted tour, ending with the Bishop's treasury. He was shown many *objets d'art* and pictures, including two Cranachs and an Orpheus by Savery.

[p.32] From this chamber I returned to the Bishop's room and sat talking to him for half an hour about the objects I had just seen. His Grace apologised for his illness having made it impossible for him to accompany me and promised to do so when he was on his feet again; but since Our Lord God had visited illness upon him, he was forced to be as patient as Job and sit still. He much preferred to suffer and watch his bishopric and subjects flourish than to be well and watch his bishopric suffer.

The Bishop also asked me what I had seen that would suit Duke Wilhelm's intentions, to which I replied that various birds and [p.33] the Orpheus by Savery would not displease His Highness. His Grace responded that he would gladly lend the birds, but that he was greatly attached to the painting. My reply was that if the Bishop were reluctant to part with the painting, he [Hainhofer] would neglect to mention it, for it is said: One does not desire what one is ignorant of. His Highness would not ask for something he knew nothing about. His Grace liked the suggestion and wanted to consider the matter further. Moreover, he announced that he and I would soon dine together, but while dinner was being prepared he would have me escorted back to my room to rest from the afternoon' exertions.

Half an hour later the trumpet was sounded in the courtyard to summon us to dinner and I was fetched to table. His Grace was brought to table in a wheelchair and taken straight to a particular place on one side. Below me on the third side sat Petrus Stephartius, Dr. Theol., Vice-Chancellor and Professor at the University of Ingolstadt, Royal Bavarian Councillor, Councillor of Eichstätt, provost at the Church of the Holy Apostles in Cologne, canon in Liège and proto-notary of the Holy See, an old man, *cui dulcis matricula senectutis est ipsius bona conscientia*.

The fourth side of the table was empty and the meal was served from that side. A jester stood there who provided the entertainment.

His Grace helped me to all the dishes himself; he drank only one of the three wines that were served, otherwise only boiled water. This he drank frequently throughout the meal since he thought to combat gout in this manner. . .

[p.34] The silver table service — serving dishes, jugs, salt cellars, beakers, court goblets, bowls, sweet dishes are all of a different manufacture to those used in my room, for there everything was round and fully gilded, whereas in my room much was rectangular and only the rims and ornamentation gilded. . .

The main topics of conversation at table were the Emperor, the conquest of Prague, the late King Henry IV of France, his wise régime and deeds, the Duke of Savoy and his intentions, England, Italy, for His Royal Grace had visited all these places, had travelled widely as a young man, and during his travels had cultivated a policy of few words, open ears and an open mind, journeying not as spiders but bees. . .

We also conversed about the weather, plants, herbs, and many entertaining matters, for Stephartius is a gifted philosopher and linguist, as is Grace, so that we exercised our minds in various languages.

When, after much good conversation, washing of hands and saying of prayers, we had risen from table, I was accompanied to my chamber and one and a half hours later fetched again to His Grace. He spent one to two hours looking at almost half my album, took several small painted objects out of a desk, amongst others a miniature likeness of himself [p.37]; this His Grace presented to me. However, the portrait bears no resemblance to his present, haggard, feverish appearance. . .

Hereupon the Bishop apologised for not dining with me as he was unwell, starting to shiver from cold again and afraid of catching a fever. [p.39] Moreover, he had to observe his diet and be moderate in food and drink. He said he would order other people to wait on me and told me to feel free to express my wishes.

When we had bidden each other goodnight, I was escorted back to my room. In the evening I was kept company by a canon who was Dean in Liège and also held a prebend in Ingolstadt; in addition by the Rector of Ingolstadt University, a French baron and also a Bavarian nobleman. Because the Rector and the canon did not speak German well we conversed mainly in French and Dutch and talked about the wars in the Netherlands.

On the following day, the Bishop sent for Hainhofer at seven o'clock, showed him his gold coins, and then, tiring, withdrew arranging for him to be shown round the main buildings in the town of Eichstätt.

Right next to the former residence is the cathedral church, an extremely old building, in the back choir of which the first Bishop of Eichstätt, a man of royal English blood called Willibald, lies buried under the altar. In the church are various altars and monuments in Eichstätt stone dedicated to bishops, although none of great artistic merit. The canons are buried in a special place and are now allowed to hang up brief epitaphs; and the large, handsome, silver altar [mentioned earlier amongst the treasures, pp.31–32] is to be placed in this church.

This same morning and afternoon I requested the Bishop's gracious permission to depart, so that I might arrive back in Munich well before Whitsun as the journey there was fourteen long miles. Moreover, I had to return from there to Augsburg where business matters were piling up as a result of my absence.

However, His Grace consistently replied: 'If I do not care for a person I soon dismiss him. That I keep you here should serve as a sign of the great affection in which I hold His Highness (as your master) and you yourself for your pleasant nature. Because, sir, you are in such a hurry to depart I shall have you delivered to Munich before the feast days; besides which, I have nothing more of service to His Highness but a few birds, which I shall give you on the way. Furthermore, if you think it a wise move, I shall give the fine picture of Orpheus (for which I paid 200 imperial thaler) to take with you at once. I only request you to tell His Highness how deeply attached I am to it and ask to have it copied at his expense.'

Hainhofer had his final audience at six o'clock on the morning of 20 May. They discussed politics, and the Bishop expressed his reluctance to join 'one or other' league, even at the request of the Markgraf of Ansbach. More treasures were displayed.

And because I once more requested permission to depart, His Grace gave me a letter for His Highness in Bavaria and asked me to convey to the Duke friendly greetings, his wish to be a good neighbour and [p.48] for Wilhelm prosperity. Further, the Bishop asked me to represent him to the Duke in the best possible light and win the latter's favour for him, promising that if he truly perceived I had carried out this office he would reward me, and that I could rest assured I would find a gracious lord not only in the Duke of Bavaria and other princes but in him as well. Moreover, he would be grateful if I could initiate correspondence between himself and the Duke of Florence because he himself exchanged the occasional letter with the Grand-Duchess's brother, Archduke Ferdinand of Graz. He expressed his pleasure at having made my acquaintance on this occasion because I had enjoyed the friendship and favour of so many princes and lords before him. He went on to say that if I would help him decorate his new great hall, above all by searching for a fine, large and unusual set of antlers, for which he would gladly pay 1000 thaler, he would repay me handsomely for my time and trouble.

His Grace invited me to return, especially during the hunting season, and said I would always be a welcome guest and assured of his gracious favour and good wishes. For all this I gave humble and fervent thanks, offering my humble but willing service and wishing the Bishop good health. I kissed his hand and took my leave.

His Grace leant against the chair with bared head, unable to take a single step, until I had made my last bow at the end of the audience chamber. Then once again he charged me with greetings to His Highness and had me escorted to my chamber to dine on diverse fish. At twelve o'clock the drawbridge was lowered and I was allowed to depart, having distributed tokens of my thanks to the silver waiter, guard, kitchen, cellar and stable staff and whoever else had served me. . .

2. Letters and documents relating to the *Hortus Eystettensis*[2]

15th April 1611
Decree by Nürnberg City Council
And since His Princely Grace [the Bishop of Eichstätt] desires, in a different letter according to which Basilius Besler, apothecary, ordered his flowers to be drawn and engraved on copper and also commissioned several citizens of the town to this end – that someone be assigned to the same Besler so that he has assistance in time of need, at the expense of My Lords (of the Council), and since it also occurred that his Princely Grace allowed it to be known, through Besler, that a certain quantity of wheat and corn was available in the Bishopric of Eichstätt, which His Grace is willing to provide cheaply to the guild of bakers; . . . so G. C. Volckamer has been required to give assistance to Basilius.
Hampe II 2388

11th/12th June 1611
Hainhofer to Philipp II of Pommerania-Stettin
Herewith I once again send a couple of pages from My Lord the Bishop of Eichstätt's book of flowers, on which work has started, so that Your Royal Highness can see what sort of format it will be published in. A proper, illuminated copy will be well worth the trouble of seeing; a white [plain] one, however, will not be so enjoyable because of the lack of description of the colours. This bishop's greatest pastime are his flowers, birds, gold and precious jewels and a handsome, large, many-forked set of antlers.
Doering: Beziehungen, S.154

25th June 1611
Hainhofer to Duke Philipp of Pommerania-Stettin
When his flower book is ready the Bishop will send Your Royal Grace an illuminated copy.
Doering: Beziehungen, S.150

10th August 1611
Hainhofer to Philipp
The Eichstätt flower book is coming along very well; it will be a fine work.
Doering: Beziehungen, S.174

---

2. The sources of these extracts are cited as follows:
   Hampe: Th. Hampe, *Nürnberger Ratsverlasse über Kunst und Künstler*, II (1571–1618), Leipzig, 1904.
   O. Doering, *Des Augsburger Patriciers Philipp Hainhofer Beziehungen zum Herzog Philipp II von Pommem-Stettin*, Vienna, 1894.
   R. Gobiet, *Die Briefwechsel zwischen Philipp Hainhofer und Herzog August d. J. von Braunschweig-Wolfenbüttel*, Munich, 1984.

27th June 1612
Hainhofer to Philipp
Work on the parchment leaves [flower paintings?] has come to a stop for the time being as Daniel Herzog of Nürnberg is working on the Lord Bishop of Eichstätt's flower book.
Doering: Beziehungen, S.235

8th August 1612
Hainhofer to Philipp
I wanted to send Your Royal Grace a May vase [engraving of May flowers], which Gärtner in Nürnberg, who is working on the Bishop of Eichstätt's book, has done for the lord Markgraf at Ansbach, who has taken the copper plate for himself. I shall see whether I can obtain an illuminated copy from the artist, who is well known to me, so that each flower can be seen in its natural colours.
Doering: Beziehungen 98, S.239

31st October 1612
Decree by Nürnberg City Council
The attention of Basilius Peßler, apothecary, should be drawn to the statement and apology made by Balthasar Caimox in the Chancellery, since the learned gentleman [Basilius] expressed concern in front of the municipal court as to how the Council in Frankfurt would react to such an action.
Hampe II 2520

20th November 1612
Decree by Nürnberg City Council
In line with the reservations expressed by the most learned lords of the municipal court in Frankfurt, Basilius Peßler shall be ordered to bring his complaint against the council in Frankfurt in a formal supplication and to hand in the same. Then he will be informed in writing, so that the security he has paid is reimbursed; but, furthermore, following his [Besler's] statement Balthasar Caimox should be called to account.
Hampe II 2528

22nd July 1613
Decree by Nürnberg City Council
Since Basilius Peßler has offered a copy of the Eichstätt Florilegium to my lords, it should be returned to him and he himself told to procure an illuminated one for my lords; a suitable arrangement will be reached about payment.
Hampe II 2587

12th October 1613

Minutes of the Eichstätt Chapter

Adam Werdenstein shall be instructed by order of the Chapter to deliver 25 copies of the first correct and perfect edition of the said book on the garden at a chapter meeting.

Staatsarchiv Nürnberg, Eichstätter Archivalien Nr 1098

1st/11th December 1613

Hainhofer to Herzog August

[In Italian] The garden of Eichstätt is on sale here, it is a beautiful book with illustrations to the life of every sort of flower in order of the seasons of the year. The book costs 35 florins. It is a beautiful book and fit for a library. If your Highness commands one of me, I will buy it. An illuminated copy will cost 500 florins.

Gobiet 26

18th December 1613

Hainhofer to Philipp

If he desires an illuminated copy of the flower book I could procure one for him but it will cost 500 florins.

Doering: Beziehungen 113, S.253

12th/22nd January 1614

Duke August to Hainhofer

You can also send me the Eichstätt plant book from the Frankfurt fair. I understand from your letter that the illuminated one is very expensive, namely 500 florins. Please be so kind as to tell me whether I have read the sum correctly or whether it should only be 50. Should it only cost 50 I would prefer the illuminated one; if it costs the greater sum I desire it only the way it is printed.

Gobiet 31

2nd/12th February 1614

Hainhofer to Herzog August

[In Italian] The Eichstätt book in natural colours costs not 50 but 500 florins, and is not to be had without great effort, since every sheet needs [1½ florins?] to colour, and if your Highness sees the copy, you will see how much labour is involved.

Gobiet 33

6th March 1614

Hainhofer to Herzog August

[In Italian] I have already many books for your Highness, including the herbal . . . and will send them either by carrier or those who go to the [Frankfurt] Fair.

Gobiet 39

4/14 June 1614

Herzog August to Hainhofer

I would like to have the plant book bound; would you please be so kind as to write me your thoughts on how it is most fittingly to be done. I think the leaves must be raised in the middle, in the way geographical plates are usually bound.

Gobiet 70

24th September/4th October 1614

Herzog August to Hainhofer

I understand from your letter that the copy of old *Teuerdank* is not complete version. In the Eichstätt herbal the copper engraving is missing of the Classis Hyberná that belongs on A4; on it is engraved 'Helleborus niger legitimus': and 'Leucoium bulbosum' in the fourth section. Please send same to me rolled up and wrapped, so that it is not creased.

Gobiet 97

6th/16th October 1614

Hainhofer to Herzog August

[In Italian] I will write . . . for the missing leaf of the Eichstätt herbal.

Gobiet 99

22nd October 1614

Hainhofer to Herzog August

The defect in the herbal was the fault of the bookbinder who bound the leaves in the wrong places, so it is not necessary to send them.

Gobiet 102

20th/30th October 1614

Hainhofer to Herzog August

[In Italian] I send the missing leaf of the florilegium.

Gobiet 105

14th October 1615

Herzog August to Hainhofer

Would you be so kind as to send a copy of the *Hortus Eystettensis* of the big herbarium, unbound but handsomely wrapped, at the first opportunity to my brother-in-law, the Count of Oldenburg and Delmenhorst, and charge it to my account.

Gobiet 187

5th November 1615

Hainhofer to Herzog August

[In Italian] I will also buy the Hortus Eystettensis for your cousin.

Gobiet 193

2nd/12th November 1615

Hainhofer to Herzog August

[In Italian] The Florilegium of Eichstätt, which costs 32 florins, I will send in box no.14.

Gobiet 194

2nd December 1615

Herzog August to Hainhofer

The two boxes, numbers 13 and 14, have arrived; I immediately sent number 14 unopened to Delmenhorst since I noticed it contained only the *Hortus E.*

Gobiet 200

11th January 1617

Duke August to Hainhofer

Please send a copy of the Eichstätt plant book, well wrapped, to us here by the next post; I want to have it for a good friend. I expect the tax will have decreased still more. As soon as it arrives the money will be sent.

Gobiet 192

9th February 1617

Hainhofer to Herzog August

I have bought one copy of the Eichstätt flower book and sent it to be stitched. It contained several omissions, so I wrote to Besler and as soon as he sends me them I shall likewise have the leaves sewn in immediately and send Your Princely Grace the book. This book is increasing rather than decreasing in value since few copies were printed and no more than 3 are said to be left.

Gobiet 296

15th March 1617
Herzog August to Hainhofer

At the first opportunity will you send me the *Gnadenpfennige* and the herbal.

Gobiet 301

6th April 1617
Hainhofer to Herzog August

I have had the flower book bound even though it was very defective in places and I had trouble obtaining the missing leaves even through the authorities, although 100 book binders, including leading ones, who have not seen a complete copy could certainly not collate this book; in addition no more than two copies are said to be still available for sale, so it can be considered to have increased in value. This copy costs 48 florins bound and had it not already been destined for Your Princely Grace, Dr Mathiol would have bought it for Dr Faber in Rome. I will give this copy for Your Royal Grace to the first carrier and send it to Nürnberg.

Gobiet 304.

3rd/13th April 1617
Hainhofer to Herzog August

[In Italian] I will send box no. 15 containing the Eichstätt flower book via Nürnberg to our Forstenhauser.

Gobiet 305

17th/27th April 1617
Hainhofer to Herzog August

From Forstenhauser I have received a letter saying that the box containing the Eichstätt flower book has aready passed through Nürnberg so it will arrive soon.

Gobiet 306

3rd May 1617
Herzog August to Hainhofer

I have received the *Hortus Eystettensis*; the binding is very fine.

Gobiet 307

30th January 1630
Herzog August to Hainhofer

Please be so good as to send me a further copy of the *Aloe America* of which Wolfgang Kilian did the engraving on 14th May 1628, for I have given away the previous one.

Gobiet 991

11th/21st February 1630
Hainhofer to Herzog August

I shall request another copy of the *Aloe Americae* from Wolfgang Kilian and send it with the next courier. . . .

Gobiet 993

16th/26th February 1630
Hainhofer to Herzog August

Enclosed are the flowers you desired. Kilian is engraving another; when it is ready I shall send that as well.

Gobiet 995

23/2 July/August 1646
Hainhofer to Herzog August

I have herbals by various authors and have the Florilegium Eystentense illuminated *ad vivum*, which I presented to the most esteemed Princely Grace of Pomerania for his library in 1617. God knows who has inherited the library, as unfortunately the entire ancient house of Pomerania has fallen like a flower and withered like grass.

Gobiet 1481

3. 'Extractus': summary of documents in the Eichstätt archives[3]

Extract from the Acts of the Eichstätt chapter relating to the 'Hortus Eystettensis', to be found in the Archives (file 3) under the heading 'Housekeeping and Management'.

(1) That the most reverend Bishop of Eichstätt and Prince of the Holy Roman Empire, Johannes Conrad von Gemmingen, graciously commanded Basilius Besler, Citizen and Apothecary of Nürnberg to produce the Hortus Eystettensis and publish it under his, Besler's name, but to dedicate the same to him, the Bishop, is shown by Besler's memorandum of the year 1612.
(2) It further appears from the 'Brief Statement and information' in the Acts that of the copperplates for the same, amounting to 370 pieces in all, 50 were prepared in Augsburg at the order of the aforementioned Bishop while he was still alive, and not less from the fact that even by then a total of 7500 fl. was advanced by the Bishop for this work.

(3) In the interval, the aforementioned Bishop died, the work being then scarcely half completed. Ibidem.
(4) Notwithstanding, it was continued and at the first impression 300 copies printed, and permission was graciously granted by Bishop Conrad's successor, the most reverend Bishop, Prince and Lord Johannes Christoph von Westerstetten, to Besler, not only to leave the plates for the Eichstätt Garden-book in his hands for 4 years but also to use them for his own purposes. According to 'Contract de Anno 1615'.
(5) On the other hand, Besler bound himself, when this time had elapsed to return the copperplates, without loss or damage. Stated in 'Recognition de Anno eodem, Mense Febr.'.
(6) But since Besler by fraud and guile sought his own advantage with this Garden-book there has been no further involvement in this place in any other edition. See 'sign D. 1615'.
(7) Regardless of which a second impression was nonetheless printed by Besler. See 'Promemoria Anno 1615, N. 5 et 25'.
(8) According to Besler's letter, the Lords [of the Council] of Nürnberg purchased from him, Besler, an illuminated copy of the Eichstätt Garden-book in the year 1617, and [he] has been obliged (NB) to pay 500 fl. 'pro labore' on sundry illuminated Garden-books, so notwithstanding (NB) the sole copy illuminated as aforesaid and sent by Besler to Eichstätt is said to have cost 500 fl. to illuminate according to account, 23 September 1613.

3. This document is translated from the transcript made in 1751 by Johann Georg Starckmann at the request of Christoph Jakob Trew, then planning an historic account of the *Hortus Eystettensis*, 'historia huius operis'. Full details of this are given in H.E., 93, 132ff. The text used here is that printed by Keunecke in H.E., 112–4.

(9) That, after Besler's death in 1629, a final settlement between here and Besler's heirs and in particular the return of the copperplates was required, is shown by the letter sent to Nürnberg to this effect. Equally, it also shows that the Besler heirs have kept the unsold copies and the money from those sold in their hands and (NB) have hitherto given no account of the separate dedication copies. Furthermore, that in the name of his Princely Grace Johann Christoph and of the then Chapter in 1630 a recapitulation was written, on account of the copperplates hitherto retained at Nürnberg, refusing permission etc.

(10) All quoted up to here and one further point is to be seen in the relevant Acta, next to which is enclosed a 'Complete Account' entitled 'Short Statement and Information of the dealings and controversies over the Eichstätt Garden-book, first between His Princely Grace and Basilius Besler, citizen and apothecary of Nürnberg, regarding the copperplates withheld by Besler and other matters still undiscussed'. Then in addition there is attached a 'Substance and Process of the Eichstätt Garden-book', compiled from extracts from the Acta and the above-mentioned 'statement', which among other matters makes clear that the sum total of the Garden-book amounted to 17,920fl.

4. Inscriptions on the drawing and engraving of the great *Agave* at Ansbach[4]

Inscription on the Drawing (plate 80)
This plant here depicted grew in a short time to 18 feet. High above the earth, as if give with white feathers, it had eighteen hundred flowers that grew in winter, before Christmas and after. All this is as I say it was. *Anno Domini* 1626.

Inscription on the engraving of the great Agave at Ansbach (plate 80)
This plant, called Aloe Americana, was bought eighteen years ago from Nürnberg to Ansbach, when it was already twelve years old. It was planted in the Markgraf's pleasure garden and carefully tended, until

4. These inscriptions appear on the drawing and engraving of the *Agave* that flowered at Ansbach in 1627, inserted in DeB.

on 4 July 1627 a thick shoot began to thrust forth, which within three months grew to a height of 24 feet, growing a further 3 feet in the winter. The surrounding shoots were 39 in all, altogether 24 feet long, but whether long or short died off, while this lower shoot had 180 buds from which the flowers came (the others, long or short, had the same shape). It reached its full bloom in winter, before and after Christmas. The size and shape of the flowers is shown at letter A. When the flower was over, there remained sometime after a growth like a cucumber, without any seeds, as shown at letter B. It is easy to imagine that, if it had not been prevented by unremitting and multifarious bad weather, in summer and winter, from growing and flowering further, it would have reached an even greater maturity. Actum, 14 May 1628. *Wolfgang Kilian sculpsit.*

APPENDIX B

Record of colour variants

1. Colour variations in individual copies

	Subject	Copies	Colour		Subject	Copies	Colour
1ii	Syringa flore cæruleo	W BSB B2 M Sp BNP BNM	blue	15ii (cont.)		N BL B1 MHN BSB T	red stamens
		N B1 DeB Er Eich	slate			BNP BNM	yellow stamens
		BL O V MHN T	magenta grey	16ii	Auricula Ursi flore purpureo	DeB	magenta shaded dark blue, yellow centre
		S	yellow			BL LU	lilac
5i	Poma flore multiplici	N Eich B1 V BNM T	white			N B1 T BSB MHN	crimson, yellow centre (T white),
		DeB BL O MHN	white, touch of pink			S BNM	
		M BNP S	red, highlighted white (S yellow)			BNP Eich	BNP green, Eich all red)
						W M	purple, brown centre
5ii	Lychnis viscosa	DeB	deep bright pink			B2 V	dull blue
		BL B1 B2 Eich Er N O V BNM T	magenta			O	yellow (transposed from 16iii)
		M BNP	orange red	21ii	Lunaria graeca annua	N B2 B BL O Eich T BSB S MHN LV BR BNM	pink-lilac
		S	bright pink, yellow centres			B1 W DeB V BN	blue
6iii	Amigdala cum flore	DeB	white	26ii	Ranunculus flore globoso	DeB N	yellow flowers, little shading
		N BL O B1 B2 BSB BNP T	white, shaded crimson (BNP orange)			BL BSB V O T BNP	yellow, shaded red
		Er2 S BNM	pink	30i &ii	Geranium Sanguinarium and Macrorhizon	DeB N T	dark/bright crimson
8i	Staphylodendron	DeB BL N Er	green/red			BL B1 O BSB MHN BNM S	pink/magenta
		B1 BSB T BNM	brown/brown			BR	magenta/pink
		B2 V O	ochre/brown — pods/			B2	pink/slate
		S	brown/red — seeds			Eich	both brick red
		BNP	grey-brown/yellow			BNP	bright blue/red, vermilion root
		MHN LU BR	green/brown				
		M	orange/green				
iv & v	Clematis Daphnoides flore purpureo (pleno)	DeB	dark blue/dark grey mauve	31i	Anemone tenuifolia flore purpureo violacea	ErZ DeB	dusty slate pink, green centre
		B1 B2 Er2 BSB O BNM V	magenta-lilac, edged crimson			BL W O V B1 BNM	slate purple, green centre
		N T	brown, edged purple			Eich B2 BNP	pink, mauve edge, purple centre
		Er	pale blue, shaded dark			BSB	white
		S (M)	violet/pink (vice versa)			MHN S LU BR	blue
		LU BR Eich	both pink (edged crimson)	39iv	Tulipa praecox	DeB	red, outlined white
13i	Guaiacana	DeB	dark purple			BL W	red, outlined yellow
		B1 BL N Erl T V BNM BNP O BSB	dark 'prussian' blue-black			N BSB Eich BNM	crimson-purple, outlined white
		M Eich B2 LU	blue			MHN S	brick red, outlined yellow-white
		MHN	green-brown			O V	red & white stripes, outlined yellow
		MC	green, pink sepals				
		S	pink, green sepals				
15i	Squammata seu dentaria maior	DeB BL N O MHN BSB	brown root/pink stems & flowers	39v	Tulipa praecox	DeB	variegated red-brick, white tips
		B1 BNM	grey root/beige, right stem only pink (lilac)			N BSB BNM	purple shaded to yellow, ditto
		B2 T V	ochre/brown-pink stems & flowers			W B1 T O V BR	crimson/pink, shaded to yellow, ditto
		BNP	all pale brown, except stems, blue-green			BNP S MHN	white, shaded red on ribs, ditto
		Er	all pink	42iii	Muscari flore obsoleto nigro	DeB	purple, shaded to pink-brown, green tips
		Eich	magenta, except flowers, yellow			BL O V T B2	dark/light blue, green-brown tips
		M OS	green, pink flowers			B1 BSB	yellow, green tips
		S	green, yellow flowers			S LU	grey, blue tips
		BR	brown, yellow stems & flowers			BNP	uniform pink-brown
		LU	uncoloured			MHN	uniform olive-brown
15ii	Dens canis flore albo	DeB	green stamens			BR	yellow & lilac, red tips

	Subject	Copies	Colour		Subject	Copies	Colour
75i	Tulipa miniata flore pleno	DeB	red	113ii	Viola Martia Canina	BL	ultramarine
		BL N O W T BNP	orange, shaded red			DeB V	cobalt, ultramarine shading
		B1	crimson			B1 B2 T N	slate, outlined crimson
		B2 Er2 BNM	pale vermilion, shaded darker			BNM BNP	pale magenta, shaded purple
		S	yellow				
75ii	Tulipa nivea oris purpurescente	DeB BNP S BR	'leaf' below flower, green	113iii	Gentianella minima verna flore caeruleo	BL	grey
		N T BL W Er2 O B1 B2	'petal' below flower, pink			DeB V	lilac
		V LV MHN BNM				O T BNP BNM S N	strong mid-blue (correct)
						LU	yellow
82	Corona Imperialis florum classe duplici	DeB	light/dark brown	115iv	Cyanus flore purpureo arvensis	DeB	lilac
		BL	yellow/brown			N BL W Erl V	grey
		W B1 B2	ochre/brown			T	chalk blue
		O BNM	grey/brown (seed-pod)			BNP MHN S MC	carmine red
		N T V Eich	dark brown/white	115vi	Cyanus arvensis flore variegato	DeB BL W N Erl	white, crimson centre
		BNP	brown/yellow			B1 B2	white, blue centre
		MHN	light brown			Eich	white, purple centre
						BNP	white, yellow & red centre
83i	Lilium Persicum	DeB	dark lilac			MHN S MC	pink, red centre
		N B1 BSB BL O W S MHN BNM	slate purple	134i	Asparagus domesticus	N T BL V BNM O	upper & lower berries red
		ErZ	purple & maroon			DeB B1 B2 LU MHN	lower berries only red
		BNP	dark pink			Sp	17 red berries at foot only
83ii	Iris tuberosa	DeB	prussian blue			S BR	all berries buff
		BL B2 BSB	ultramarine	144i	Alcea Syriaca	DeB	ultramarine, maroon centres
		N O BNM	slate (as 83i)			W	mid-blue, crimson centres
		B1	dark brown, thin blue edge			N B1 B2 O V BNM	dark lilac, vermilion centres
		BNP S MHN	pale green, black turnovers			Eich, Ellw	pale grey-pink, red centres
90iii & iv	Palma Christi	DeB	both crimson			MHN BNP	red, darker red centres
		B1	crimson/pink			S	vermilion
		BL N T BNM	white/crimson	149i	Herba Paris	DeB BL	black, crimson bracts
		W	white/white			V	blue-black, brown bracts
		MHN LU BR BNP S	red or magenta/white			N B1 B2 W O BNM Ellw MHN	blue-black, green bracts
91ii	Scabiosa silvestris	DeB N T	light/dark blue			S LU	red, green bracts
		B1 B2 BL W BNP BNM S V LU	all strong mid-blue	155i	Pyramidalis Lutetiana	DeB	grey
94i	Rosa Damascena flore pleno	DeB N T W B1 B2 BL BSB	white			BL N B1 W BNM	slate
		MC S BR	red			MHN BNP LU	blue
100iii	Melissa Fuchsii	DeB	blue inside flower			S	green
		BL V BNP BNM	magenta-crimson	165i	Leucoium flore rubro pleno	DeB N B2 W BNM LU BNP	plain red
		MHN J	pink			BL O V	red, yellow to green centres
103i	Paeonia mas flore purpureo	DeB BNP	stigmata uniform green			S Ellw	vermilion, small green centres
		BL O V	pale green, red tip	167i	Leucoium flore purpureo pleno	DeB B1 B2 W V BNM	pink, shaded grey & dark purple
		N T BNM	pale green, white highlights, red tip			BL	lilac
		MHN S	white, red tip			BNP Ellw	carmine red, darker shading
		DeB MHN BNP	red			MHN	purple-red
		N V B1 B2 BL V W	orange (seeds)			S	magenta, yellow centres
		S	yellow	170i	Aquileia variegata	BL W B2 Erl MHN S	blue & magenta
107i	Paeonia Bizanthina maior	DeB MHN N T	dark red			N B1 V BNM Eich DeB OS	blue only (DeB yellow, white, blue buds, others all green)
		B1 BL Erl V S BNP BNM	bright red			LU Ellw	
		W	orange				
107ii	Paeonia Bizanthina minor	DeB	dark red	170ii	Aquileia flore rubro	BL	dark red, much white
		BL N Erl T BNP MHN S BNM	crimson			DeB N B2 Eich Erl	red-crimson
107v	Borago flore cæruleo	DeB	lilac			BNM MHN	magenta-purple
		N BL W Erl V	grey			BNP Ellw	bright red
		T	chalk blue				
		BNP MHN S MC	carmine				
111i	Nymphaea alba maior	DeB B1	brown wash, green detail on root				
		BL V BNM	ochre wash, no detail				
		N Eich O	green wash, no detail				
		BNP	grey wash, no detail				

Subject	Copies	Colour	Subject	Copies	Colour
170ii (Cont.)			224iii (Cont.)		
	O OS	orange red			orange shading
	BR LU	pale red		S	yellow flowers
170iii Aquileia Stellata Rubescens	BL DeB N B2 W O BNM	magenta		MC	red flowers
	B1 W OS S MHN	dark red	232i Calamintha montana praestantior	DeB B2 BL BNM	lilac (DeB shaded yellow)
	BNP Ellw LU BR	pink/bright red		W N BNP	crimson, white highlighting (BNP none)
175iii Consolida regalis multiplicato violaceo flore	DeB BNM	variegated light/dark lilac-grey		B1	yellow
	N BL BNP	pink-grey, blue shading, green stamens		MC	blue
	V B1	grey, white stamens	232ii Calamintha montana vulgaris	DeB B1 W N BNM	white, touches of pink
	W	steel-blue, white stamens		BL	lilac
	OS	crimson, yellow stamens		BNP	crimson
185ii Iris Bulbosa pallido colore	N W BNM DeB BL B1	all yellow		MC	yellow
	MHN S	blue & yellow		MHN	blue
	BNP BK	lilac	239i Horminum hortense	N BL	white/grey
187i Lilium montanum flore purpurescente punctatum	V W B1 DeB BNM	purple, buff corm		B1 B2 DeB	white/blue
	N BL O	lilac, shaded green, yellow corm		BNP MC	pale blue/blue
	BNP	bright magenta, brown/white corm		MHN	pink/blue
192 Dracontium maius	DeB	dark brick	245i Hyosciamus vulgaris	B1 DeB N B2 BNM	pale yellow (DeB veined darker)
	B1 BL BNP LU	crimson-dark red		BL MHN	pale yellow, veined mauve
	Eich	pink, veined crimson		BN	dark pink, stamens yellow
	N W B2	dark brown			
	MHN	black, purple highlights	245ii Hyosciamus albus	N BL BNM (BNP)	white (BNP yellow stamens)
197i Orchis Serapias secunda Dodonaei	DeB N W	crimson-pink, ochre-green centre, above; black & orange-brown, yellow edge, below		DeB MHN	pale yellow
			258i Scabiosa hispanica maior	N	magenta, green centres
	BL V Eich B2 BNM	same, but no yellow, above or below		BL MHN BNM	brown, green centres
	BNP	plain lilac, yellow edge		BNP	brown, white centres
	MHN	brown, yellow centres, pink petals		DeB B1 B2	green (DeB two shades)
	S LU	brown, red centres, yellow petals	265ii Blattaria flore luteo	N BL V BNM BNP	lemon
				DeB MHN	orange
197ii Iris Bulbosa lutea mixta	DeB	flower veined brown	271i Lotus Urbana	N B1 DeB	slate-blue
	BL W N BNP OS	flower veined red		BL BNM BNP MHN MC	pale blue
198ii Polygonatum angustifoliis	N B1 DeB BNM	black berries	272i Trifolium Bituminosum odoratum	DeB BL B1 N MC	red
	W BL MHN BNP	green berries		B2 MHN BNM BR	grey/blue
	B2 S	red berries			
202ii Iris bulbosa variegata	DeB	greenish/orange	272ii Trifolium Bituminosum inodoratum	DeB BL	slate, shaded dark blue
	BL BNP	'prussian' blue/orange		N W B2 BNM MC	white, shaded pink blue
	W B2 LU	yellow/orange		MHN	blue
202iii Iris bulbosa mixta	DeB (W)	'prussian' (W pale) blue/purple-orange	274i Acanthus spinosus	N DeB	purple/green
	BL BNM MHN	lilac/orange		W B1	brown/green
	N LU	lilac/blue		BL B2	crimson/green
	BNP	blue/yellow-white		BNP MHN LU	red/white
206iii Calendula lutea medioruffa	N B1 BL BNM	green centre spot in crimson stamens		BNM	blue-grey/green
	DeB W MHN	no green spot	275iii Acanthus Levis	N B2 BNM	green/dull lilac
	BNP	centre all green		W Er BL	green/crimson
222i Malva rosea multiplex flore incarnato	DeB	dark crimson		BNP MHN	red/white
	BL N Eich BNM B2	magenta, crimson outline & shading		DeB	green/strong chestnut
	W S BNP	straw, heavy crimson shading to brick	280i Carduus Sphaerocephalus	DeB	green flowers, bright crimson tips
	MHN	dark pink		BL MHN LU N	green flowers, reddish brown tips
224iii Marubium Creticum angustifolium	BL	pink flowers		B1 B2	green, brick red
	DeB N MHN	white flowers		Eich NP BNM	green throughout, white/blue tips
	OS	lilac flowers,	288i Papaver spinosum	B2 MC S	red (all others yellow)
			289i Papaver corniculatum liteum	DeB N W BNM	emerald/olive leaves
				BL BNP MHN	plain green (MHN blue-green)

	Subject	Copies	Colour		Subject	Copies	Colour
293i	Papaver multiplex album ovis rubicundis	DeB	white, much pale lilac detail	335i	Iasminum Indicum, seu flos mirabilis peruanus	N W DeB T B2 B1 LU BNM	L flower plain yellow
		BL N B1 B2 W BNM	white, strongly detailed crimson			BL	1 red petal
		BNP MHN	white, much pink-red detail			MHN	striped red & yellow
293iii	Papaver flore pleno argentei coloris	DeB	light buff, heavy purple shadows			BNM N W O DeB T	R flower symmetric pink & yellow
		BL N W BNM	lilac, long darker lining strokes			BL BNP	R flower asymmetric irregularly pink and striped
		B2 BNP MHN	bluish detail			MHN	
298i	Sedum vulgare	N BL B1 B2 W BNM	pink edges to stem leaves	340i	Tabacum latifolium	N W O T B1	crimson edges, white inner, green centre
		DeB BNP MHN	stem leaves plain green			DeB BL B2 BNM	pink edge (line only), white inner, yellow centre
315iii	Caryophyllus purpureus flore multiplici laciniato	DeB	bright dark purple/green			BNP MHN	red/pink, green centre
		BL	magenta/green			S	yellow edges, pink centre
		N W BNM	dull pink/grey	345i	Papas Peruanorum	DeB BL	red berries, green at edge
		BNP MHN S	crimson-red			N O	yellow berries, green at edge
321i	Melanzana fructu pallido	N DeB BL S T W L Sp	all fruit purple			MHN	plain yellow
		B2 BNP BNM	all fruit yellow			BNP	red, white highlights
		MHN	purple left, yellow right	348i	Cyclamen serotinum foliis hederaeis	DeB	dark 'ring' on leaves
322i	Balsamina foemina	DeB Eich BNM	all purple			N W BL B1 B2 T OS	white 'ring'
		BL B1 B2 N W Er T BR	1 purple / 2 green (upper 3 buds)			BNP S LU BR	leaves plain
		BNP MHN LU	all green	359	Ficus Indica Eystettensis	BL DeB Er W	orange posts, red tub
		MC OS S	all blue/orange/red			N BSB	red posts, orange tub
		Ellw	1 green / 2 white			Eich B1 BNM	both red, posts shaded darker
322ii	Momordica fructa luteo rubescente	OS MC	blue flowers (all rest pale yellow)			O B2/OS/S BNP	orange, shaded red/brown/black
329i,ii	Piper indicum maximum	DeB	bud left red, concealed fruit right green, all others vice versa			V MHN	blue posts and tub
						BL N T DeB Er W B1 B2 O Eich	stem and lower fruit brown
334i	Iasminum Indicum, flore rubro & variegato	N T BNM BNP	top left, both facing flowers mixed			MHN BR LU	stem only brown
		BL B2	L white, R red			BNP S	all green
		DeB W Er MHN LU	L red, R white	360iii	Fructus Opuntiae dimidio dissectus	DeB N B1 B2 T	yellow left, red centre
		O S	both flowers white			BL W Eich BNM BNP BR	red left, yellow centre
						MHN	all red

2. Colour variations in individual subjects

4.iii *Summitates Piceae* is rendered very differently, N, BL pink at tip graduating to a green base, DeB, Er green round the edge with pink central highlight. V & O correspond with N, BL, but are russet, not pink, LU darker still; B1 & B2 are mainly green, with pink tip only, BR all green. ErZ, Eich are mainly pink, Eich yellow at base. BSB is brown above, pink below, half and half. BNP is dark grey at edges, highlighted pinker grey in centre, tip of left cone brown.

10–12. Considerable variation in the treatment and colour of both flowers and seed pods of various forms of Cytisus. Flowers vary from canary to orange. DeB tends to latter, BL, N & W have crimson shading, BSB green. The treatment of seeds and pods in *Colutea Vesicaria* varies: some (BL, DeB, U, LU, B) have greeny yellow to brown pods with a darker interior. Others show seeds distinctly and in a different colour: W (ochre/yellow), Eich (plain/brown), S (pink/brown), B1 & B2 (plain/black); V has ochre pods with blue interior and brown seeds.

15. The variety in the colour of *Squammata* (15 i) is perhaps explained by the difficulty of preserving its colour out of the ground. On the other hand, *Pulmonaria* (15 iv) is common, and the fact that no two copies have the flowers coloured alike is perhaps evidence of familiarity.

19. The variety of colour in the outer petals of *Cariophyllata montana* (19 ii) from grey through magenta to red and brown again suggests that no coloured model was used. Eich (yellow out, white in) is eccentric.

20. *Orobus Pannonicus* (20 ii) is given with all flowers diversely coloured pink, white and blue in DeB. Others are blue and pink (? without highlighting).

22. *Anemone*: considerable variety in treatment of stamens.

38. *Hyacinthus orientalis variegatus* (ii) and *mixtus* (iii) vary between pale grey to pink to crimson, between flowers. (O, T and MHN are mainly blue).

41. *Hyacinthus stellatus peruanus* (i) is rendered with complete uniformity, as indicated by single flower drawn in ErZ, except S which lacks blue.

44. *Ornithogalum Neapolitanum* has varying yellow at centres of flowers, from none (N) to overall (BN).

48. The pansies, *viola flammea maior violaceo colore mixta* (ii) and *pallido colore* (iii), vary considerably. The upper leaves of (ii) are uniformly

	blue, but the amount of yellow varies from little (B2, Eich, BSB, O, TB, MHN) to large (DeB, BL, N, ErZ). This is repeated for heartsease (289 ii, iii).
50.	The stem of *Colchicum vernum* (iii) varies from pale grey (W, B1, B2, Er) to lilac or purple (N, BN, LU, BR). T, S, MHN, BSB are green above, white below. The flower varies from pale blue grey (Er) to dark blue (W, BSB); median is slate (DeB), shaded red (N, B2) or purple (BL). T, exceptionally, has green, MHN strong violet.
50–66.	The scales of *Narcissus* bulbs (brown without, white within) are shown with varying degrees of fidelity to the engraved lines, as also *Hyacinthus* (36–49). The later copies do not show this contrast, using plain brown overall in various shades, with some highlighting.
67–79.	The tulips are shown with great uniformity in the early copies, with very fine detail and a surprising number of colours. This pattern become diminished and conventionalized in the later copies. DeB shows, eccentrically, strong maroon on *T. lutea lituris aurea* (70 i) and *T. lutea maculis rubens* (71 i); it also shows the top leaf of *T. nivea oris purpurescente* as green, uniquely among the early copies (75 ii). T has an eccentric alternation of lilac and white petals in *T. diversis coloris* (70 iii).
95–8.	Red roses are notably deep and strong in N, DeB and BL. Other copies (W, V, BSB, O, B2) have more white highlights (brown on white roses). Eich has a brick red, B1 between red and magenta, without detail. *R. provincialis flore incarnato pleno* (95 iii) is mauve (T pink).
117–18.	The various types of *Chamaeiris* are very differently coloured, *latifolia biflora* ranging from solid purple (DeB) to grey (N), and *flavo et purpurescente* from mauve and pale yellow (B2) to crimson (T).
119–24.	Irises are equally variable, particularly *I. Calcedonica latifolia*, where painters reacted differently to the superb engraving, some repeating detail in paint (N, grey, Eich, grey and crimson), others tinting only (T, purple, O and S flesh, LU blue). BL and DeB have green tongues in the middle of lower petals, others yellow. ErZ detail corresponds exactly with engraving.
134.	*Asparagus domesticus*. The distribution of red and green berries differs. N, T, DeB, BL, OSB have red above and below, LU, B1 and B2 in lower half only. Sp has a few only in lower centre; S has all buff.
135.	*Arbor vitae*. There are differences in colour and distribution of seed pods.
138,140.	Both V and B1 have a brownish cracked surface to green and orange, suggesting that a transparent varnish, perhaps albumen-based, was used to heighten the sheen. This also found in S.
142.	*Mala Armeniaca*. Considerable variation in colour, from white shaded crimson to ochre yellow shaded purple, from fruit to fruit; the early copies show the greater variety.
144.	*Alcea Syriaca*. Varies from blue (W, DeB) to pale grey-pink (Eich), with centres varying from maroon to vermilion Ellw has only one flower tinted. S and MHN are plain red.
161.	*Linum sylvestre* is mainly bright blue (N, V, B2); BL is, improbably, grey with purple detail; DeB has dark blue edges, paler inside, to white centres.
170.	The variously coloured *Aquilegia* suggests that a familiar plant was coloured from knowledge, not a model.
174.	The elaborate structure of *Melanthium Hispanicum maius* (Nigella) flowers is differently treated, the large inner stamens being brown (W, BL), brick red (DeB) or green with brown tips (O, V), the outer purple (W, BL), brick red (DeB) or green (O, V).
175–8.	The many *Consolidae* show many slight colour variations.
181.	The corm of *Martagon Imperiale Moschatum*, generally yellow, is brown to yellow in BN, yellow outlined brown in O, brown-pink MHN, and light and dark brown in OS. This is repeated in various degrees in other lilies (183–91).
182.	The flower varies from lilac to pink, filaments green or white, anthers yellow, ochre or orange.
183–4.	The buds are generally crimson, paler only in B2, MHN.
215.	MHN alone shows *Aster Atticus maior flore coeruleo* (iii) with variegated flowers.
224.	Slight but very various differences appear in *Malva crispa* (i), again suggesting nature rather than a model.
247.	DeB has *Hypericon* (iii) a realistic chrome-orange (*H. maculatum*); others beige to pale yellow.
253.	*Flos Armerius variegatus* (iii) is very variously coloured, DeB alone showing red and white only, others showing up to three different shades of pink to crimson.
279.	*Cinera cum flore* represented a considerable challenge to colorists. N, W show the flower blue only, as do later copies (S, LU, O, BR), B1, OS, blue-grey with paler stripe, Eich, B2 introducing pink. BNP and MHN alone have a buff stripe. DeB and BL have the flower purple at the top, turning white tinted red at the base. The tips of the bracts are painted white (those in N crimson, in OS vermilion). The leaves are shown as dark green (upper sides) and green bice (lower).
280.	*Carduus Sphoerocephalus* (i) was a similar challenge. The head is generally green (MOS red, BN white) and the tips crimson (B1, DeB, BL, N) green-blue (BNP, MHN) or orange (OS).
282.	*Eringium maritimum* (i) is incorrectly shown as green (N, DeB, B1) but rightly blue-green (BL, B2). BL, W, T, show flowers correctly rust-red, BR lilac. It is unlikely that this was based on nature, even at Eichstätt, but rather on a drawing.
290–3.	The large poppies show considerable variation, not in colour, but in detail, all finely worked.
293ii.	*Papaver flore pleno argentei coloris* is generally lilac (B2 bluish), with darker long lining strokes. DeB uniquely is light buff, with heavy purple shadows and base.
296.	*Lupinus Sativus maior* (i) is correctly given by DeB and MHN. BL, N, W, B1 are slate to mauve, perhaps an example of the use of the same pigment changing with time, but *Lupinus Sylvestris* (ii) is uniformly and correctly blue.
300 ff.	The arched intertwining of species begins here continuing intermittently to '323' (322).
302.	*Soldanella Marina* is generally tinted dark red, quite wrongly, in early copies (DeB, BL, B1, B2, N, T), perhaps due to the absence of a natural model (see 282). Later copies are paler (MHN, S, BR).
308–16.	The many examples of *Caryophyllus* are treated with varying degrees of detail. Generally the reds are darker and bluer and the white less blue than MHN. In N *Caryophyllus maximus* (310 i) has two long white stamens painted on to the four main flowers (cf 315).
320.	*Poma amoris fructu rubro*. The fruit at lower left of stem is variously red (N, B2) or green (DeB, BL); that above is bicoloured, differently.
'322'.	*Solanum Pomiferum* (i). Unripe fruit are differently tinted (BNP uniquely gives them as blue).
324–31.	The peppers again have small differences of detail, most notably at 329 (v.s.). The red in most copies (N, DeB, BL, W, B1, B2, Eich) is without highlights. T and BN show fruit all brown at 325, and MC divides them slate left and orange right at 327 and 329.
344.	The two Sorghums vary from crimson to straw and white to ochre, every copy being slightly different.
348.	The leaves of *Cyclamen serotinum foliis hederaceis* are usually painted a uniform green bice, shaded darker where engraving indicates with a white painted hatched ring variously nearer to edge or middle of leaf. In DeB this hatched ring is dark green, not white.
349–52.	Some variation in colouring of variants of *Colchicum*.

APPENDIX C
Engravers' and illuminators' signatures

Part	BL (Mack 1615)	N (Schneider) (Æ: G. Schneider 1613)	T/L (Schneider) (H: Mack 1614)	B1	V	M	BSB	O	Eng W. Kilian J. Leypolt
Title	(Mack 1615)	(Schneider)	(Schneider)						
Portrait	(Mack 1614)	(Schneider 1613)	(Schneider 1613)						
1	—	G.S.							
2	GM	GM	GM						
3	GM								
4	—								
5	GM								
6	GM		⊬						
7	GM								
8	GM				GM				
9	GM								
10	GM								
11	GM								DC
12	GM								GH
13	GM	G.S.	G.S.						
14	GM	G.S.	G.S.						
15	GM								
16	GM								
17	GM								
18	GM								
19	GM	G.S.	G.S.						
20	Geo: Mack								
21	GM								
22	GM		G.S.						
23	GM		G.S.						
24	GM	GM							
25	GM								
26	GM								
27	GM								
28	—	G.S.	G.S.						
29	GM	GM							pj
30	GM	G.S.				MF			
31	GM	GM	GM			HSF		JL	
32	GM	GM	GM					JL	
33	GM								
34	GM								
35	GM								
36	GM								
37	GM								
38	GM								
39	GM								
40	GM								
41	GM								
42	GM								
43	GM								
44	GM								
45	GM	G.S.	G.S.						
46	GM								
47	GM	GM							
48	GM		G.S.						
49	GM		G.S.						
50	GMA								
51	GM								
52	GM		GM						
53	GM	G.S.	G.S.						
54	GM	GM	GM						
55	G	GM	GM						JL
56	GM		G.S.L.						
57	Geo. Mack		—						
58	GM	G.S.	G.S.						
59	GM								
60	GM		G.S.						SVR
61	GM	GM	GM						
62	GMA	GM	GM						
63	GM	GM	GM						
64	GM								

Part	BL	N	T/L	B1	V	M	BSB	O	Eng
65	GM								
66	Georg Mack		DR		GM				JL
67	GM								JL
68	GM	GM1							
69	Georg Mack								JL
70	Georg Mack								
71	GM								
72	GM		GM						
73	Geo Mack								JL
74	GM								
75	Ge: M	GM/1613	GM						
76	GM	GS	GS						
77	GM	GM	GM						JL
78	GM/2 Marti. 1614								(VH)
79	GM						M/46615		
80	GM/1614		DR						
81	GM/1614		Georg Schneider						
82	GM	Georg Schneider							R. Custodis
83	GM								
84	GM		GS						
85									
86	GMA/1614	GS							
87	GMack/1614								
88	GM	GS.L:	GS.L:					Hans VZ. Jul	(J. Leypolt)
89	GML	GS							(VH)
90	GM								
91	Gö A								WK
92	Ge: MA	GM					MF		
93	GM					GM			
94	GM				GM				(VH)
95	GM								
96	GM	GM							
97	GM	GS	GS						
98	Ieorg MA				GM				
99	GM	GS	GM						
100	GM								
101	GM								
102	GM	GS	G.S.I.		GM				RC
103	G Mack		G.S.II.						
104	GM								
105	GM								
106	GM	G Mack	G Mack						
107	GM								
108	Geo: M:				GM				(W. Kilian)
109	Georg Mack/ 1614		G.S.						
110	GM								
111	GM	Ge: Mack	GM						
112	Georg Mack								
113									
114	GMª/1614								
115	Georg Mack								
116	GMA								
117	Georg MA/ 1614				GM		IM		
118	Georg Mack 1614				GM				
119	Georg Mack								
120	Georg Mack	Georg Mack 1613	Georg Mack 1613						(S. Raeven)
121	Georg Mack	G.S.	G.S.		GM				
122	GML		GM						

	BL	N	T/L	B1	V	M	BSB	O	Eng
123	GM			GM					
124	GMA	G.S.	G.S.						
125	Georg Mack/ 1614			GM					
126	GMA								
127	Georg Mack/ 1614	GM	GM						
128									
129	GM	GM	GM						
130	GM			GM					
131	Ge: MA								
132	GM							DK	
133	GML	G.S.	G.S.					DK	
134	Georg Mack Illumi: 1614								
135	GM		GS						
136	GM		GS						
137	GM		GS						
138	GM	H.S.	GS						W. Kilian
139	GM	G.S.	GS						
140	GM	G.S.	GS	GM					JL
141	GM	G.S.	GS						
142	GM		GS	GM					
143	GM		GS	GM					
144	—	G.S.			GM				
145	—								
146	GM				IN				
147	—								JL
148	GM			GM					
149	GM								
150	GM			GM			MF		
151	GM	G.S.							
152	GM		DH	GM			MF		
153	GM	H.S.	GH	GM		IN			GH
154	GM								(S. Rauen)
155	GM								
156	GML	GM							
157	GML								
158	GML	G.S.							
159	GM								
160	GM	GM		GM					
161	GM								
162	GM								
163	GM	G.S.							
164	GM								
165	GM	GM		GM					
166	GM								
167	GM			GM					
168	GM	G.S.							JL
169	GM	G.S.		G Mack 1615					
170	Georg Mack			GM					JL
171	GM								FH
172	Georg Mack	G.S.							
173	GM								S Rav
174	GM								
175	GM	MB							DK
176	Geo: MA								
177	GM								
178	GML	D.R.							
179	—	GM							
180	GM								
181	GM			G					
182	missing						MF		
183	GML				RC	R			RC
184	Georg MA	Georg Mack 1613		GMA					(P. Isselburg)
185	GM								
186	Geo: MA			GMA					
187	GM 1614								
188	GML	GM							
189	GM	G.S.							
190	Geor: MA								
191	GMA	G.S.							

	BL	N	T/L	B1	V	M	BSB	O	Eng
192	GM								
193	—	GM			RC				RC
194	Georg Mack								(S. Rauen)
195	Georg Mack								
196	GM								
197	GM		G. Mack 1613						JL
198	GM								
199	GM								
200	GML	GM							
201	Georg Mack	G.S.							
202	Georg Mack	G.S.							
203	GM	G.S.							
204	Geor Mack								
205	GM								(H. Ulrich)
206	Geo: Mack								
207	GM	G.S.			GMA				
208	—				GM				
209	GMA								
210	GM		gold G.S.						
211	G Mack								
212	Geo: Mack 1614								
213	Georg Mack Illu: 1614	G.S.							
214	Georg Mack.								
215	GML				GAH				HL
216	GM	G.S.			GMA				
217	—	G.S.							
218	GMA								
219	GM								
220	GMA								
221	Georg Mack 1614								
222	GMA								
223	Georg Mack.								JL
224	GM								
225	GMA 1614								
226	Jeorg Mack 1614								
227	Jeorg Mack Illum: 1614								
228	Georg Mack								
229	GMA	G.S.							
230	GM	G.S.			GM				
231	—								
232	GMA								
233	GMA	G.S.							
234	GML			G.M. Seni:					
235	GM								
236	GM	G.S.							
237	GMA	GM							
238	GMA								
239	GM								
240	GM								JL
241	GM								
242	—								R. Custodis
243	—				GM				DK
244	GMA				GM9				
245	—	G.S.							DK
246	GM								
247	GMA								
248	GM								(H. Ulrich)
249	GM								
250	GM	GM							
251	GMA								
252	GM								
253	GM								
254	Georg Mack	GM							
255	—								
256	G Mack								
257	G Mack	G.S.							
258	GMA:	G.S.							
259	Georg Mack 1614								

	BL	N	T/L	B1	V	M	BSB	O	Eng		BL	N	T/L	B1	V	M	BSB	O	Eng
260	Georg Mack									315	GM								
261	GMA	G.S.						DK		316	Georg Mack								JL
262	GMA									317	G Mack 1614								
263	GM									318	G Mack								
264	GM	GM	GM							319	GM	G.S.	G.S.					MFürstin	Pj
265	GM									320	GML	G.S.							
266	GMA									321	GM	G.M.	GM						
267	GMA							(S. Rauen)		322	GM	G.S.						HSF	
268	Ie Mack									323	GM	G.S.							
269	GMA	G.S.								324									
270	—							(S. Rauen)		325									
271	GML	G.S.								326									
272	Ieorg Mack 1614	G.S.								327									
273	GMA									328									
274	GM	G.S.								329									
275	GML									330									
276	GMA									331									
277	GML	G.S.								332	GML	G.S.	G.S.						
278	Ge:Mack	G.S.								333	GM	G.S.	G.S.						pj
279	GM	G.S.								334	G Mack. 1614.								JL
280	Georg Mack 1614	G.S.								335	Geo: Mack 1614.								
281	Georg Mack 1614									336	Georg MA	G.S.	G.S.				RTAF		
282	Geo: Mack									337	Georg Mack 1614.								
283	Georg: Mack									338	GML	G.S.	G.S.						
284	GM	G.S.						DK		339	GM								
285	GMA									340	Georg: MA	MB							(VH)
286	GMA									341	Geo: MA.								(VH)
287	GMA									342	G.MA	GM							
288	GMA	G.S.								343	GM				GM				
289	GM						MF			344	GM								JL
290	Georg MA								JL	345	Georg Mack 1614		G.S.						
291	G. Mack							(W. Kilian)		346	Georg Mack	GMI	GM					HTF 1671	pj
292	GM																		
293	GM								pj	347	GM	G.S.	GS.						
294	GMA									348	G Mack 1614								
295	G Mack.									349	GM		Georg Mack						pj
296	GM																		
297	GML									350	Geo:MA	G.S.	G.S.						
298	Georg Mack	GM						(S. Rauen)		351	GMA	G.S.	G.S.						JL
299	GM									352	GM	G.S.	G.S.						
300	Georg Mack 1614.									353	GM	DR	G.S.						
301	GM									354	—	G.S.	G.S.					MF	
302	GM									355	GMA								JL
303	GM	G.S.								356	GMA								(VH)
304	G Mack			GH						357	GMA		GM	GM					JL
305	—								JL	358	G. Mack: 1614		G.S.					GM	(H. Ulrich)
306	G Mack	G.S.								359	—		(Schneider 1613)	(Schneider 1613)				(M. Fürstin 1677)	JL
307	GMA									360	—							MF	JL
308	GM							(R. Custodis)		361	Georg Mack 1614.								
309	Georg Mack 1614									362	Geo: Mack 1614								
310	Georg Mack 1614								JL	363	G Mack								
311	G Mack							R. Custodis		364	Ge : M								
312	—									365	GM								
313	G Mack									366	GM								
314	G Mack							(W. Kilian)		367	GMA								

[75]

APPENDIX D
Analysis of Schedel Kalendarium

The later damage and repairs make it impossible to establish the original form of the Kalendarium. It looks as if unbound sheets (some already numbered) had been grouped together for binding, after which the leaves were consecutively foliated. The blank leaves indicate that the book was not 'complete' at this stage. More material was added, in part (presumably) drawn direct on to blank leaves, part by inserting new leaves, and part by pasting in smaller pieces of paper or cut out drawings and engravings. This seems to have been a continuous process; the fact that most of the material from the *Hortus Eystettensis* appears on versos suggests that it may be a relatively late addition.

The folio numbers are those of the original full sequence (inserted leaves are indicated as 'A', 'B'). The subjects are given within quotation marks, if so captioned; if not the common or Beslerian name is used, as appropriate. The degree of resemblance to the *Hortus Eystettensis* is hard to convey, but if 'close' then the drawing seems to correspond with at least part of the engraved image. Details of added material (including cut out prints), watermarks, and other peculiarities are listed in the last column.

Folio	Caption / Subject	Resemblance to HE Close	Remote	Notes
1				
2				
3				
4	} missing			
5				
6				
7				
8				
9	'Petasites fl. albo', 'Salicis flos' (2)	365 iii		
10	'Pulsatilla', 'Consiligo, sive Hell[		366 iii	
11	'Aconitum'		365 ii	
12	'Narcissus 8. Matthiol. vulgo Schneetropfchen'		361	arms of Nürnberg w/m
13	'Narcissus VII Matthioli leucoium'			
14	'Schneetropfchen'	361 v		
15	'Crocus vernus'		367 iii	
16	(more, untitled)			
17	blank			
18	blank			
18	blank			
19	blank			
20	blank			
21	blank			Slip with hyacinthus botryoides pasted to verso
22				
23	} missing			
24				
25	'Hyacinthus stellatus', 'Narcisssus 7 Matthioli', 'Hyacinthus fl. caerul.', 'Daphnoides', 'Hyacinthus fl. cinericio'			
26	'Hyancinthus botryoides', 'Muscari fl. flavo, obsoleto', 'Iris', 'Anemone'	42 i	45 i, ii	
27	'Hepatica nobilis fl. incarn.' & 'fl. caerul.'			verso: slip with yellow tulip (77v)
28	'Iris Tuberosa', 'Pseudonarcissus'	83 ii		verso: slip with pinks 28v
29	Hyacinthus flore albo, violaceo, 'Tulipa', Pseudohelleborus		39 iii, ii, v	Tulipa praecox, yellow
30	Hyacinthus Orientalis maximus, Primula veris fl. dupl.'		36 ii	
31	'Adonis vernalis'	364 iii		
32	pulmonaria		151 v	
33	helleborus niger	362 i		
34	'Dens caninus', snowdrop		15 ii	foliated '43'
35	'Pulsatilla' and Primula auricula	366 iii	16 iii	
36	Ranunculus Illyricus, fl. albo dup.	26 i	27 i	
37	'Narcissus', Fritillaria', 'Viola fl. dup.'	62 ii, 56 i	18 i	
38	'Viola' (3), Daphne mezereum (?)	18 iii, iv, v	367 ii	
39	'Petasites' and Pulmonaria	365 i	15 iv	verso: slip, more daffodils
40	Daffodils		56 iii	
41	blank			
42	blank			
43				
44	} missing			
45				
46	blank			
47	} missing			
48				
49	'Apprilis' part-title, 'Dens caninus'		15 ii	Auricula, heartsease on verso
50	'Flos Trinitatis' (yellow and purple heartsease)		289 ii, iii	Verso: 67 iv&v, cuttings
51	'Tulipa'	67 v		Verso: 67 iii cutting
52	'Tulipa' ('silvestris v.' added in another hand)	70 i		verso: 67 ii cutting
53	'Anemone Pavo', verso Cheiranthus Cheiri	32 iii, 169 i		169 i close but without the deformed leaves (see f.172 below)
54	Tulips	67 i	66 ii	67 i reversed, but exact (possibly a coloured counterproof)
55	'Tulipa'		69 ii	
55A-B	Tulips (four)	69 iii(2), 71 iv	68 iii	55B: remains of pressed flower
56	'Tulipa'		75 ii	
56A	Pressed tulip flower, stem coloured	67 iv		
57	'Tulipa'		67 v	
58	'Tulipa'		74 iii	
59	'Tulipa'	70 i		
60	'Tulipa'		73 v	
60A	pressed tulip flower, stem coloured		72 iii	
61	'Tulipa'	73 i		
62	'Tulipa'	69 v		
63	'Tulipa'	71 iii		
64	'Tulipa'	70 i		
65	'Tulipa'		73 v	
66	'Tulipa'		71 iv	
67	'Tulipa'		77 i	
68	'Tulipa'		78 i	
69	'Tulipa'		73 ii	
70	Tulip		68 iii	
71	'Iris'	120 ii		
72	'Tulipa' (2)		74 v, 73 v	
73	'Iris Florentina', tulip	120 ii	74 v	
74	'Tulipa', 'Ornithogalum Arabicum'		74 v, 91 i	
75	Tulip		77 i	
76	'Tulipa'		71 i	
76A	Tulips	68 v, 68 i		added leaf, w/m arms of Nürnberg
77	Tulip		74 v	
78	'Tulipa'	67 i		
79	'Tulipa'	69 i		
80	'Tulipa'	68 i		
81	Tulip		77 i	L flower only
82	Tulip		72 i	R flower only

Folio	Caption / Subject	Resemblance to HE Close	Remote	Notes
83	'Paeonia'		100 i	
84	'Paeonia'		100 i	
85	Paeonia	100 i		
86	Corona Imperialis		82	sideways, on double fold
87	Tulip and Bellis	112 i	74 v	
87a	Pressed tulip flower, stem painted		74 iii	
88	'Tulipa' (2)		70 v,?	Yellow tulip with ragged petals, shaded green, not in HE
88A	pressed tulip flower, stem painted		73 iii	
89	Tulip		67 v	as f.51 in style
90	'Tulipa' and 'Anemone'	31 iii?		
90A	Pressed tulip flower, stem painted		74 iii	inserted leaf
91	Tulip		70 i	
91A	Tulip	73 v		inserted leaf
92	Tulip		73 iv	
92A	pressed flower, stem coloured	69 i		inserted leaf
93	'Fruchlings Veilc[hen]'	18 iii, iv, v		
94	'Geranium', Fumaria, L. xylosteum, Berberis	30 iii, 25 iii, 9 ii, iii		
94A	Tulip (2)		39 iv	
95	Tulip			
96	Tulip			as f.90
97	'Ranunculus globosus', 'Tragopogon'	161 iii	26 ii	
97A, B	Tulips		69 iii, 71 v, 73 iii	pressed flower, inserted
98	Calendula	206 ii		
99	Tulip			as f.95
100	Anemone (3)	27 iii, 3 ii, 29 i		
101	'Pulmonaria', 'vinca pervinca', 'Chaemaecissus'	237 ii, 7 iv&v	125 iii	
102	Anemone, Lychnis, ?		27 iii, 20 iii	
103	Syringa	1 ii		
104	Buxus, Arbor Iudae, vinca minor	1 i, 7 iii	3 i	
105	Cytisus, Malus	10 iii, 5 i		
106	'Paeonia mascula'	103 i, 104 i		
107	'Paeonia mas'	same		
108	Tulipa	71 iv?		
108A	Tulip	71 iv, 78 i, 74 i		
109	Tulip	74 i		verso: 70 iv cuttings
110	'Flammula Iovis' (Thalictrum)	25 ii		verso: tulip 73 i, 75 ii?
111	Narcissus minor	54 i		
111A	Tulip (2)		as 74 i,	inserted leaf
111B	Tulip (2)		71 iv	inserted leaf
112	Tulip (cut out)	69 ii		verso: 70 iii cuttings
113	Tulip	78 iii		
114	Tulip (cut out)		? 71 i	drawing cut out, as HE fragments
114A	Tulip (4)			
115	Tulip (2)		67 i, 71 i	
115A	Tulips,			
116	Tulip (2)		75 ii	
116A	Tulip (3)			
116B	Tulip (2)			
117	blank			verso: 70 iv cuttings
118	Iris	118 i		flower only verso: 70 ii, cut
119	blank			verso: 70 iv, cutting
120	'Maÿ' title, red silk tag			verso: moth & butterfly
121				Leaves cut out
122?	'Asphodelus' (missing), Ornithogalum, hyacinthus stellatus	92 i		cut away at top; verso, insects
123	'Narcissus totus albus fl. pleno'	52 i		
	'Narcissus medio luteus Pisanus'	55 i		
	'Narcissus medio luteus'	55 ii		
124	'Narcissus', 'Simphitum' (2)	51 i, 244 i		
125	'Althaea', 'clematis'	300 i	219–20	
126	'Hyacinthus Rubellus Anglicus' & 'Hispanicus'	40 iii, 44 ii		44 ii reversed
127	'Lilium convallium'	131 ii&iii		
128	'Anemone', 'Hyacinthus stellatus', 'Calendula'	29 iv, 43 i	206 ii	
129	'Anemone' (3)	31 iii, 32 iii		
130	'Anemone', salvia		31 iii, 238 ii	
131	Sambucus, 'Leucoium', 'Ranunculus', Bellis	10 i, 112 i	167 iii, 26 iii	10 i reversed
132	Iris			
133	'Aquilegia'	171 i		
134	Hemerocallis	130 iii		verso: irises 117 ii, 118 iv
135	'Iris'	117 ii	118 iv	verso: Iris pumila
136	'Iris Anglica bulbosa'	200 iii		
137	(leaves)		201 i	verso: purple iris
138	'Iris bulbosa'			
139	Iris Calcedonica	120 i		
140	'Lychnis'			verso: slip with pansies, pinks, &c.
141	'Iris', 'Lunaria', 'Saponaria'	121 iii, 21 ii	not 261	
142	'Iris bulbosa'	200 ii		
143	'Saponaria'		not 261	verso: lily
144	Lily	85 i		
145	'Arum'	33 i		sections only, reversed
146	'Tulipa', L. xylosteum	9 i		
147	'Iris', 'Narcissus Iuncifolius polyanthus'	123 i, 66 i		123 i reversed
148	Scorsoneria latifolia		255 i	
149	Hyacinthus stellatus			
150	'Cyanus maior'	114 ii		
151	Paeonia		108 i	
152	Rosa pendulina	98 iii		
153	Bellis, &c.		22 i	verso: slip with corn-flower, chicory, &c.
154	'Geranium muscatum'			
155	same			
156	'Lilium persicum'			top of leaf torn out
157	leaf torn out			
158	'Geranium'		22 ii	
159	blank			
160	Papaver rhoeas: pieces cut from HE 'Papaver eraticum'	290 ii		
161	Iris variegata: root and leaves cut from HE	123 ii		verso: flower 123 i, root 123 ii
162	Iris pallida: flowers cut from HE	123 i		
165	White and red double peonies			cut out & pasted in
166	'Junius' title			
167				leaf cut out
168	'Digitalis'	150 ii		reversed
	verso: caterpillar, slip with cyanus maior		130 ii,	
	Lis Martagon (2)	187, i 188 i		187 i reversed
169	'Martagum'	189 i		189 i reversed
170	'Viola Mariana' (Campanula medium)	152 i, 153 i		
171				
172	Mutant wallflower			not like 169 i, but perhaps its original
173	'Centifolia'			
174	'Digitalis'		150 i	
175–6	18 individual 'Marvel of Peru' flowers	335		
177	Cnicus sativus		276 i	
178	Impatiens Balsamina	322		
179	'Gladiolus'	201 iv		
180	'Gladiolus'		202 ii	
181	'Consolida regalis', 'Carduus' (not in HE)	175 i		

Folio	Caption / Subject	Close	Remote	Notes
182 }	'Barthnelcken' (?) =	253 iii, ii		
183 }	Dianthus barbatus	253 iii, i		
184	Hemerocallis		186 i	
185	'Rosa lutea fl. pleno'	95 iii		reversed
186	'Ranunculus globosus'	26 ii		
187–8	'Sigillum Salomonis'	92 iii		
189	Symphytum officinale			
190	'Calceolus Mariae'	122 i		reversed
191	'Gelbe Roses'		95 iv	
192	Lythrum salicaria (?)		268 ii	
193	Lysimachia vulgaris		267 iii	
194	Caryophyllus	314 iii		
195	'Althaea Syriaca, seu Rethmis'		219 ii	
196	Galega officinalis	273 ii		
197	Philadelphus coronarius	2 iii		
198	Iris			no parallel
199	Lupinus luteus		296 ii	
200	Hazel nut & white dianthus			
201	White dianthus		317 iv	
202	Chrisanthemum tenuifolium		307 i	
203	blank			
204	ribes`		86 iii	
205	'Jullius' title, Centaurea			
206	missing			
207	'Papaver'	290 i		
208	'Papaver'		293	
209	'Papaver'			
210	'Thlaspi Creticum', Caryophyllus	236 iii		
211	Caryophyllus			
212	Iris fetida, 'Flos Hierosolimitanus'	124 i	254 i	
213	Calendula		206 i	
214	'Nigella', 'Hyosciamus', &c.	124 iii	245 i	verso: lilium montanum: 190
215	Acorus Calamus		125 i	
216	Salvia sclarea ('Horminum hortense', *HE*)		239 i	
217	blank, except stalk			
218	'Campanella', 'Carduus phaerolus', 'Cirsium'	277 iii	248 ii, 280 i	
219	'Ornithogalum Arabicum',		91 i	
220	'Martagum'	187 ii		
221	'Tulipa'			
222	'Iris', 'Ornithogalum Pannonicum'	197 ii	90 ii	
223	'Iris bulbosa', 'Gladiolus', 'gelbe Roses'	202 iii, 95 iv	201 iii	202 iii reversed
224	'Ranunculus Tripolitanus'	29 iv		
225	'Gladiolus'		203 ii	
226	'Genista Hispanica', 'Napellus'		11 iii, 160 ii	
227	'Blattaria'		265 iii	
228	'Acandium silvestre'		280 ii	
229	'Indianische Feigen'	360 ii&iii		
230	'Antirrhinum', 'Cervicaria', antirrhinum	157 ii, 154 i	157 i	
231	Cheiranthus, Aquilegia	167 iii	170 ?	
232	Malus	206 iii		
233	Calendula			not like 344
234	Panicum miliaceum			
235	Amaranthus maior panniculis rubris		338 i	
236–7	blank, 238 missing			
239	'Ficus informalis', 'Flos noctis'	301, 302 i		
240	Solanum pseudocapsicum 'Strichnodendron' (*HE*)		148 i	
241	'Siciliana', 'Rotes Holier'		247 i	
242	'Papas Peruanorum'		345	
243	'Lilio Narcissus'	65 i		
244	'Lagopus'		271 ii	
245	'Hysopus'		283 iii	
246	'Primula veris'	112 iv		
247	Rusans hypoglossum		128 i	
248	'Cucumis Asininus'		284 i	
249	'Orchis Serapias'	197 i		
250	'Linaria Vulgaris'		102 iii	
251	Daphne mezereum, Amaranthus tricolor	367 ii, 337 i		sketches only
252	Aquilegia, Anemone, Gladiolus (Victorialis rotunda)	170 iii, 16 v, 201 ii		sprays only
253–7	blank			
258	'September' title			
259	torn out			
260	'Semen paeoniae maris'	103 i		
261	'Semen paeoniae maris'		103 i	
262	'Indianische Rosen' (Tagetes erecta)	305 i		
263	'Stramonium vel Datura Turcorum'		343 i	
264	'Pappeln'			
265	'Pappeln'			
266	Amanitis phalloides			not in *HE*
267	Capsicum		328 i, ii	
268	Capsicum, Scabiosa	328 i	259 ii	
269–75	missing			
276	'October' title			
277	Indian corn head			
278–81	missing			
282	blank			
283	'Noũember' title			
284	missing			
285	blank			
286	missing			
287	blank			
288–9	Index A–G (H–Z missing)			

APPENDIX E
Colour Reversal

The evidence that some copies were coloured from written instructions, rather than a coloured master copy or samples, comes from the instances of 'colour reversal', the transposition of colours between two adjacent specimens on one page, or in a single multi-coloured specimen. These instances do not present a clear picture, since the colourists naturally do not err uniformly or invariably. Even from the outset, too, the colourists (with good reason, perhaps) did not always follow the verbal definitions of colour that appear in some of the captions, while in some later copies differences are due to the colourist following these indications more exactly.

The reason for supposing that these transposals are due to an error of reading, rather than careless visual examination, lies in the irregular numeration of the plants on each plate. There may be any number of plants, from one to seven; normally (but there are exceptions, e.g., plates 95–9, 114) the plant numbered 'i' is in the centre, but beyond that there is no principle of arrangement. If there are three plants, that on the right is normally numbered 'ii' and the left 'iii' (the most frequent cause of confusion). If there are more than three, the numbering normally starts at the foot and moves upwards. But there are many exceptions in both cases, and both clockwise and anti-clockwise sequences are found. What is hardly ever found is the normal 'reading' sequence, left to right, head to foot. Thus, faced with a written list of colouring instructions numbered sequentially from one, the colourist may on occasion ignore the numbers on the print and start on the left or at the head of a complex page.

The transposals noted are as follows:

16 ii & iii : O	176 ii & iii : Eich	270 i & iii : S
39 ii & iii : S	179 ii & iii : S	292 iii : LU[3]
42 ii & iii : BR	191 i & iv : BR	298 iv & v : S
72 i & v : S	196 ii & iii : MHN	315 i & ii : S
90 iii & iv : BNP, MHN, S	201 iii & iv : MHN	364 i & iii : S
117 iii & iv : BR	212 i & ii : S, MC	365 i & ii : S
121 ii & iii : S[1]	232 i & ii : B1 & MC[2]	

1. Here too the plate drawings (ErZ) have the captions transposed, which may be the beginning of a tradition of reversal. It is in any event difficult to identify the different irises, and there are many variations, as noted above (Appendix B).

2. Here B1 and MC each show one of the Calamints with a yellow flower; it may be accident or a verbal instruction minconstrued by one or the other.

3. LU here reverses the inner and outer colouration of the poppy.

APPENDIX F
Chemical analysis of modern copies

As noted earlier, the contemporary market for individual leaves of the *Hortus Eystettensis*, suitably coloured, is so strong that several complete but uncoloured copies have been broken up in recent times to supply it. In addition, two complete uncoloured copies have been (rather unsatisfactorily) coloured within the last few years, one offered for sale by Zyska & Kistner, 25–7 October 1988, lot 1706 the other by Sotheby's, June 1993. The separate leaves, which are widely available at the present time, are harder to identify, and are potentially more dangerous, the more so since since at least one authentically coloured copy (MC) has been broken up in recent times.

Considerable care has been taken to ensure that some at least of the contemporarily coloured leaves conform (to outward appearance) with those coloured in the 17th century. Others, more easily distinguished, are simply coloured by reference (or even without it) to modern photographically illustrated botanical reference books. The apparently authentic coloured copies are based on a complete set of transparencies, popularly supposed to have been made from the British Library copy. In May 1988, Messrs Bernard Quartich kindly arranged for ten of these transparencies to be made available to me, and a minute inspection revealed a number of details peculiar to the Eichstätt copy. I have been unable to ascertain when they were made, but it was more than ten years ago.

However, recent advances in the analytical technology of chemistry offer an alternative method of detecting modern coloured copies. The pigments available in the 17th century were largely organic and included no synthetic elements. The earliest datable artificial pigment, Prussian blue (ferric ferrocyanide), was only developed in the first decade of the 18th century. Most such pigments are more recent, few pre-dating 1850, and modern advances in dye-stuff technology result in frequent changes and new developments which can be dated with some precision.

In order to take advantage of these, a coloured leaf from the *Hortus Eystettensis* (plate 285, sold Reiss & Auvermann, 15–18 October 1991, lot 3549) was purchased, and, by courtesy of the Chemistry Department, University College, London, tested, using Laser Raman Spectroscopy and a polarizing microscope. The print was chosen because it contained a fairly wide range of colours, in the hope that an equally wide range of pigments would be found. This expectation proved to be well-founded, and an interesting number of different chemical components were detected; in addition, the value of different types of test and equipment was also demonstrated.

In all, five different kinds of pigment were present, in different degrees of concentration and mixture, as follows:

1. *Red*. Laser Raman Spectroscopy identified this as Quinacridone Red:

Examination under polarizing microscope revealed characteristic magenta crystals. Quinacridone Red is a synthetic organic pigment first developed in Germany in the 1930s, but not made commercially available till the 1950s in the U.S.A. It is not yet listed as a standard colour (R. Mayer, *The artist's handbook of materials and techniques*, London, 1987, para 5.2.2).

2. *Blue*. This was identified by polarizing microscope only as Phthalocyanine Blue (copper phthalocyanine, $C_{22}H_{16}N_8Cu$). This is also a synthetic organic dyestuff, developed as tinctorially stronger and more light-fast than Prussian blue or ultramarine by chemists at I.C.I. It was first marketed by I.C.I. under the name by which it is now usually known, 'Monastral Blue', at an exhibition in London in November 1935 (R. J. Gettens and G. L. Stout, *Painting materials: a short enclyclopaedia*, New York, 1966, 136–7).

3. *White*. This was analysed using a Scanning Electron Microscope, and the resultant peaks show calcium, zinc, and an unusually high level of barium. This, if not technically impossible, is most unlikely in an old pigment. As, in the present case, it is applied over red (to provide pink), and the red pigment is shown to be modern, further analysis is unnecessary.

4. *Green*. This was examined under the polarizing microscope. The blue component of this mixture was identified as (again) Phthalocyanine Blue. The yellow proved more problematic, since it could not be positively identified using either Laser Raman Spectroscopy or polarizing microscope.

5. *Yellow*. Using a Scanning Electron Microscope, the following result was achieved. The main elements are $CaSO_4$; calcium alone might be found in a traditional lake, but in combination with sulphur and barium, it must indicate an organic precipitated on to a substrate with calcium (chalk) in a formulation which, again, is extremely unlikely to have been made more than a century ago.

In summary, the presence of either red or blue pigment alone would have provided evidence of the modern date of the colouring. As the blue was used in combination with the yellow to produce the green, this also could be identified as modern. Had there been any doubt about the blue, the fact that the un-mixed pigment used on the flowers was painted after the red and with the same brush, so that there are traces, visible by microscope, of Quinacridone Red in the blue, condemns it too. In short, all the pigment analysis shows that the colour was added to the print not earlier than the 1950s.

NOTE ON THE PLATES

The plates are, of necessity, only a small selection of all the different points that distinguish 24 coloured copies of the *Hortus Eystettensis* and the three related books of drawings. Each has the consecutive number of the plate, using the *Besler Florilegium* order, the copy abbreviation and the modern Latin name of the plant depicted. Where more than one plate appears on the page, they are lettered, left to right, top to bottom. The reasons for the selection are various: most illustrate significant or unique details of colouring (these are listed in Appendix B), and, in particular, instances of colour reversal (Appendix E); others show illuminators' signatures (enlarged details of these are given in plate 77); others again simply depict the quality of botanical or pictorial rendering, notably in the De Belder and British Library copies. The plates from the Kew Kalendarium and Camerarius Florilegium illustrate the resemblance of these sources to the equivalent likenesses in the *Hortus Eystettensis*. The last three plates show the work of Abbot Johann V Müller at Bamberg, and the non-botanical masterpiece of Sebastian Schedel; and, finally, the drawing and Wolfgang Kilian's engraving of the Agave that flowered at Ansbach in the De Belder copy (the discrepancy in the dates between drawing and print two is unexplained; the inscription on the drawing appears contemporary, but the date on the print is predictably followed by later printed sources).

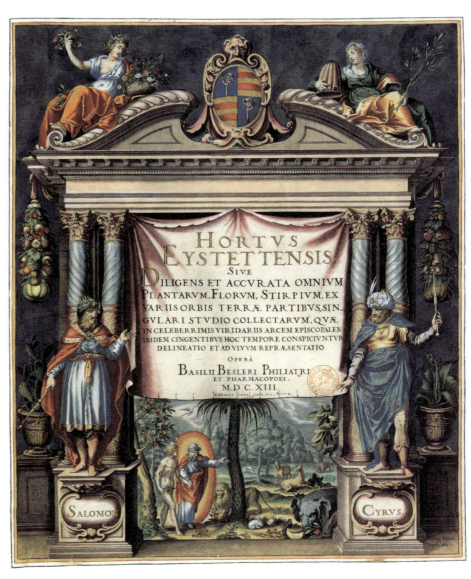

1   a N; b DeB; c W; d BNP

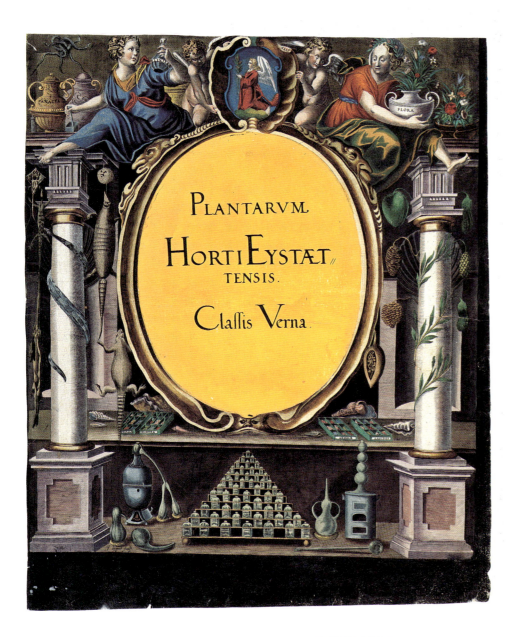
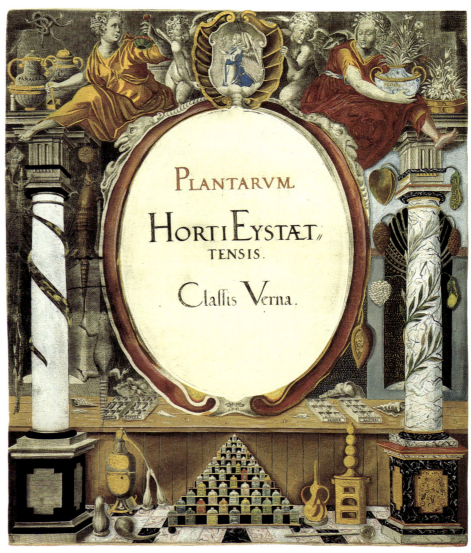

2   A DeB; B LU ('Classis Verna'); C 4 DeB; D 4 S (*Prunus cerasus, Prunus padus, Picea abies*)

8   16 A N; B T; C BL; D O (*Primula auricula* i–iii, *Corydalis bulbosa*, *Anemone nemorosa*)

10 30 A BSB; B BR (*Geranium sanguineum, G. macrorrhizum, G. tuberosum*); C 31 BSB (*Anemone coronaria, Anemone hortensis* ii & iii)

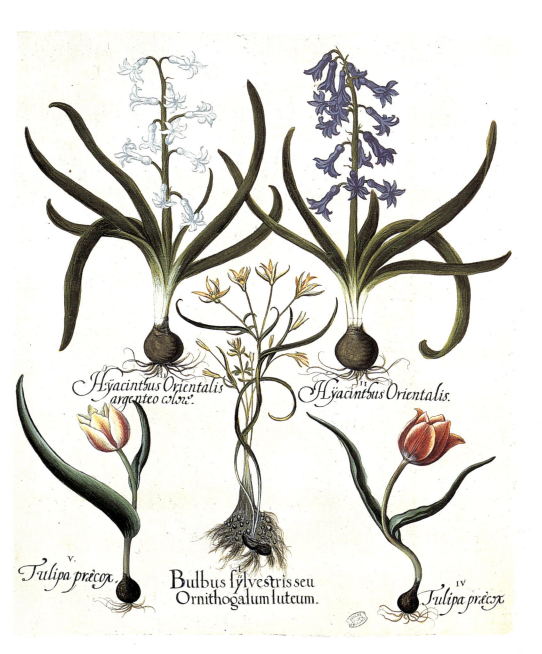
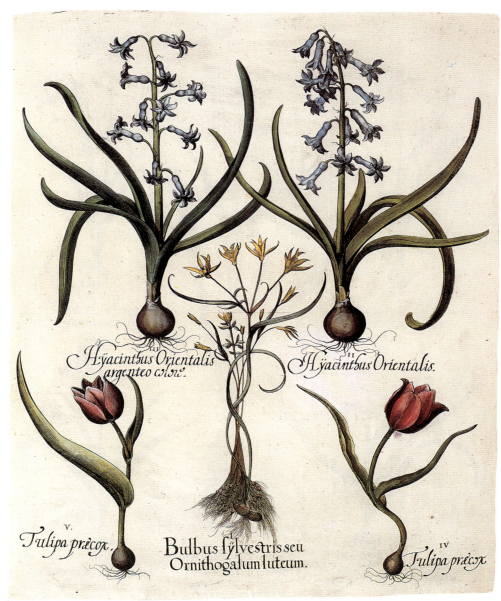

11 39 A MHN; B S; C DeB (*Gagea arvensis, Hyacinthus orientalis* ii & iii, *Tulipa praecox* iv & v), 42 DeB (*Muscari neglectum, M. moschatum* ii & iii)

14 A & B Schedel Kalendarium, ff 50, 54; C & D 67 ErZ & Er (*Tulipa*)

15 67 DeB (*Tulipa*)

16 71 BL (*Tulipa*)
17 A 72 S, B 73 DeB, C Camerarius Florilegium, f.142 (*Tulipa*)

18   75 A BL, B DeB, C B1 (*Tulipa*)

19 A 77 Er; B Schedel Kalendarium, f.114 (*Tulipa*)

20   78 A DeB, B V (*Tulipa*); C S; D 83 B1; Camerarius Florilegium, f.13 (*Lilium persicum*)

22 88 T (*Lilium album* i & iii, *L. chalcedonicum*)

23   90 DeB, S (*Ornithogalum thyrsoides, Ornithogalum narbonnense, Gymnadenia conopsea* iii & iv)

24   94 A MHN; B N; C Camerarius Florilegium, f. 77

26 95 DeB (*Rosa centifolia, R. variegata, R. provincialis, R. hemispherica*)

Iris Florentina. — Iris Calcedonica latifolia. — Iris Illyrica.

Iris latifolia maior variegata. — Iris latifolia violaceo colore maior. — Iris latifolia vulgaris cœrulea.

Iris Calcedonica

30 A & B 120 N & T; C 121 S; D Schedel Kalendarium, f.139

31 120 DeB (*Iris susiana, I. florentina, I. illyrica*)

32   122 DeB (*Cypripedium calceolus, Iris variegata* ii & iii)

34 144 DeB (*Hibiscus syriacus, Anthemis cotula, Malcomia maritima*)

38   161 DeB (*Linum perenne, Tragopogon dubius, T. porrifolius*), Camerarius Florilegium, f.54

II. *Flos Cheyri simplex medius.*   I. *Flos Cheyri maximus Eystettensis.*   III. *Flos Cheyri simplex minor.*

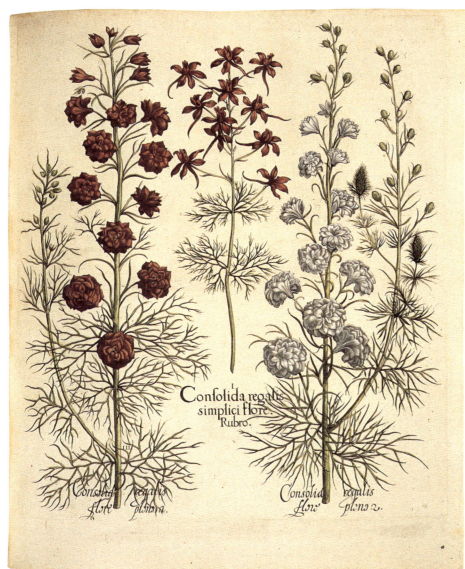

40   170 A BL, B DeB (*Aquilegia vulgaris*); C 176 Eich (*Consolida regalis*) D 179 S (*Matthiola incana, Erysimum cheiranthoides, M. tristis*)

41 174 DeB (*Nigella hispanica*, *N. sativa*, *N. damascena*)

42   181–2 DeB (*Lilium imperiale, Dianthus sylvestris, D. gratianopolis*)

45 196 A S (*Epipactis helleborine, Dactylorhiza incarnata* ii & iii, *Platanthera bifolia, Nigritella nigra, Goodyera repens*); B 197 DeB *Ophrys fuciflora, Iris xiphium* ii & iii); C 198 B2 *Iris latifolia, Polygonatum verticillatum*; D 202 DeB (*Allium ampeloprasum*); E Camerarius Florilegium, f.149

Flos Solis maior.

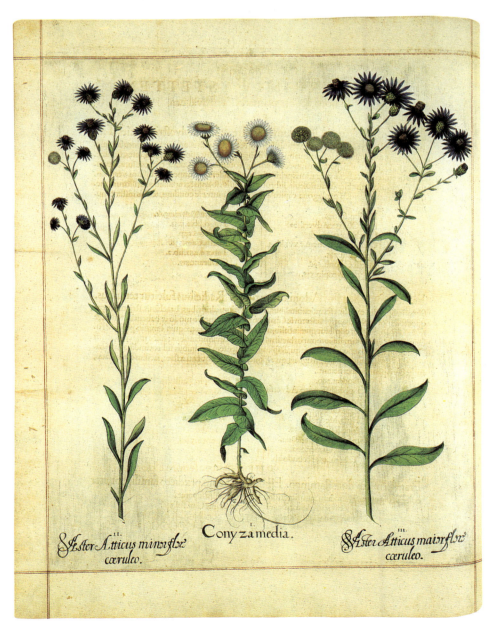

47 A 212 S, B BL (*Anaphalis margaritacea, Senecio bicolor*); C 215 V, D L (*Pulicaria dysenterica, Aster amellus* ii & iii)

Malua Rosea multiplex flo: albo.

Grosse weisse doppelte Herbstrosen,
Grosse weisse volle Pappelrosen.

Malua Rosea multiplex flore incarnato.

Ehrenrosen, mit doppelt leibfarben blumen.

49   232 A B1 (*Calamintha montana, C. grandiflora*); B BL, C 234 L (*Ocimum basilicum*)

Flos Armerius variegatus. — Flos Armerius albus. — Flos Armerius ruber.

51   264 L (*Actea spicata, Filipendula vulgaris*)

52 A 270 S (*Potentilla palustris, P. recta, P. erecta*); 272 B DeB, C BL (*Psoralea bituminosa*); D Camerarius Florilegium, f.172

53 274 DeB (*Acanthus spinosus, Myosotis sylvatica, M. arvensis*)

55 288 a B2, b BL, c MC (*Argemone mexicana, Helichrysum italicum, Achillea ptarmica*)

56 A 290 BL (*Papaver hortensis, P. rhœas*); B Schedel Kalendarium, f.207

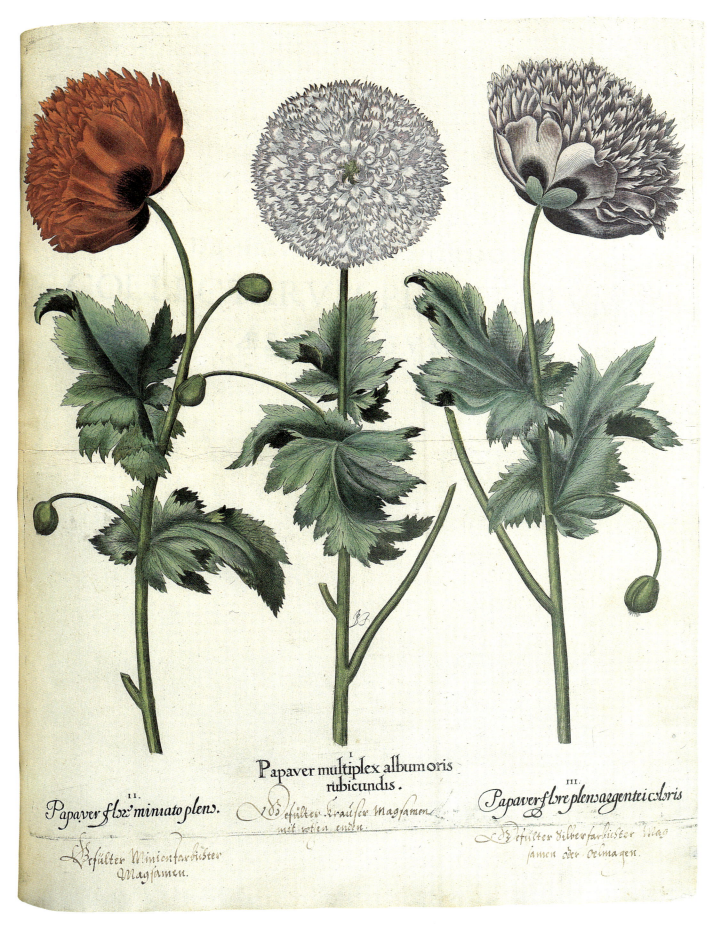

57 A 293 DeB (*Papaver hortensis*); B Camerarius Florilegium, f.172

58 294 BL (*Tropaeolum maius, Bellis sylvestris* ii & iii)

59 298 A S, B BL (*Sempervivum tectorum, Sedum villosum, S. acre, S. annuum, Lathyrus tuberosus, Pisum sativum*); C 308 BL (*Dianthus caryophyllus* i & ii, *D. deltoides*); D 309 LU (*Dianthus caryophyllus*)

II. Melo Saccharinus variegatus I. Poma amoris fructu luteo. III. Pseudocolocynthis Pomiformis

62   A 319 BL (*Lycopersicum esculentum, Cucumis melo, Citrullus colocynthis*);   B Camerarius Florilegium, f.179

Melanzana fructu pallido.

64 Schedel Kalendarium, f.178

65 323 BL (*Impatiens balsamina, Momordica charantia*)

67   335 A DeB, B BL (*Mirabilis jalapa*);   C Schedel Kalendarium, f.176

Ficus Indica Eystetten, sis ex uno folio enata luxurians.

Stachlichter Indianischer feigenbaum

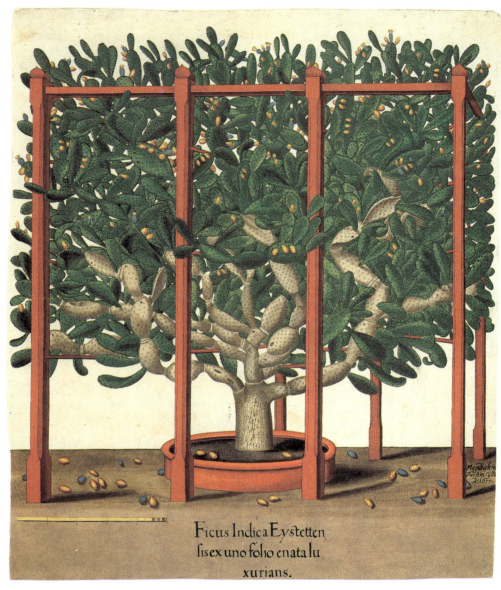

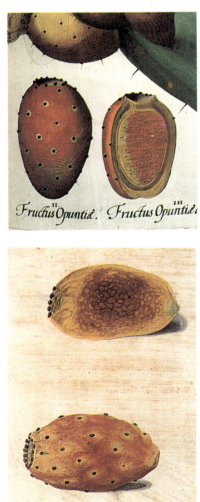

73   359 A N, B O;   C Camerarius Florilegium, f.177;   D 360 DeB (*Opuntia ficus-indica*);   E Schedel Kalendarium, f.229

74   361 A MHN, B MC, C S, D BL (*Helleborus niger*, *Galanthus nivalis* ii & iii, *Lencoium vernum* iv & v)

75 363 DeB (*Helleborus fœtidus, Crocus chrysanthus, C. angustifolius*)

76 365 A BL (*Petasites hybridus*, *Eranthis hiemalis*, *Petasites albus*); B Schedel Kalendarium, f.39

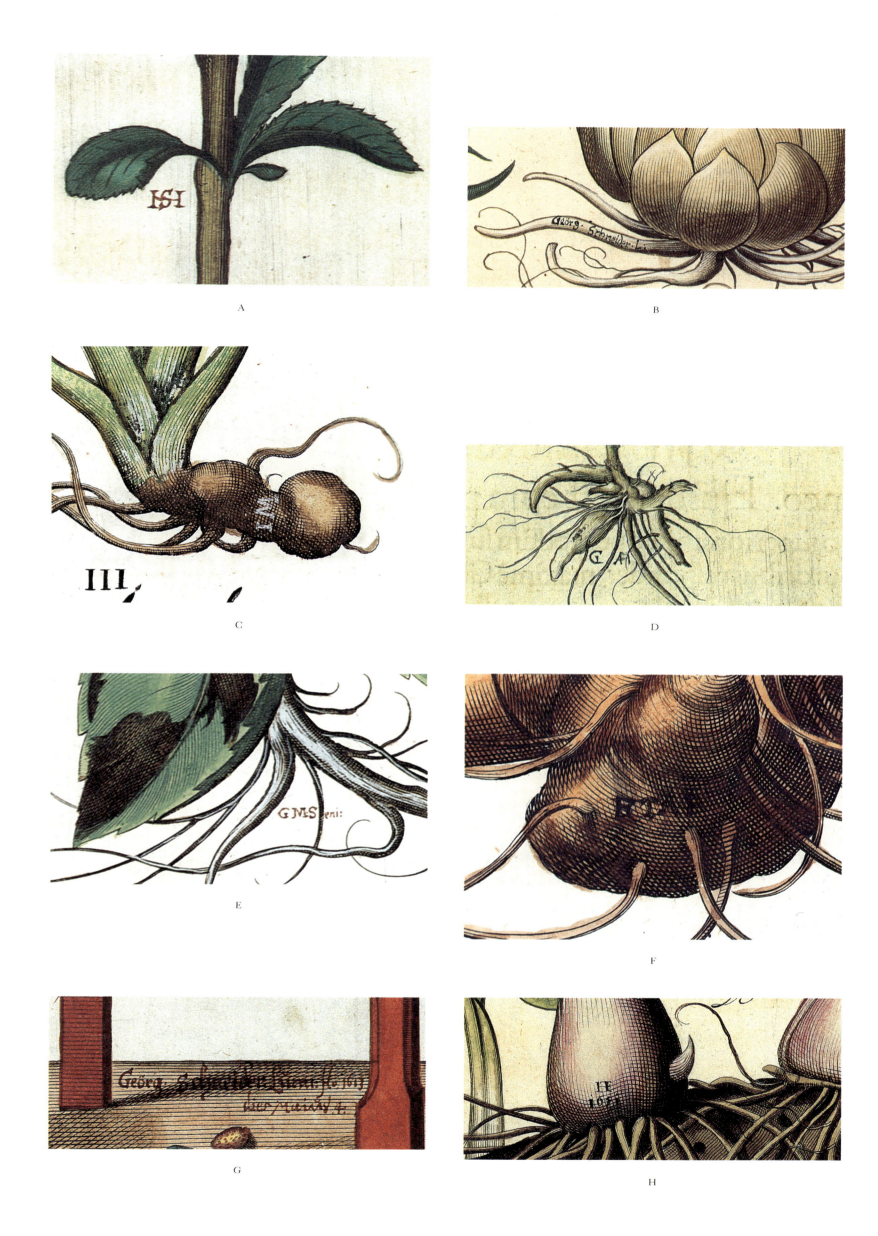

77 A 6 T; B 88 N; C 117 BNM; D 215 V; E 234 L; F 336 BNM; G 346 O; H 359 T

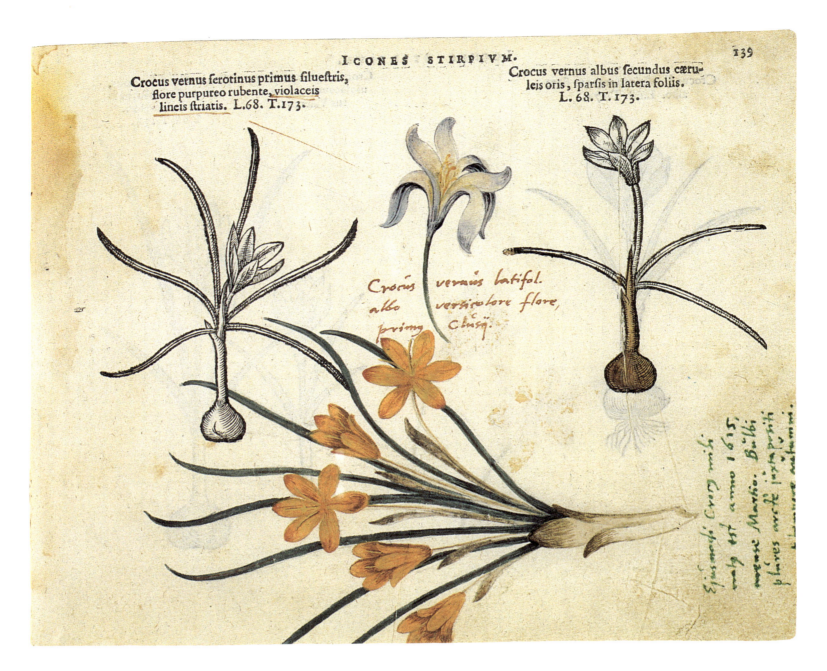

78 Bamberg, Staatsbibliothek, Lobelius *Plantarum sive stirpium icones* 1581, pp.139, 252

79 A Bamberg, Michaelskirche; B Sebastian Schedel, 'Schembartlauf', U.C.L.A., MS. 170/351

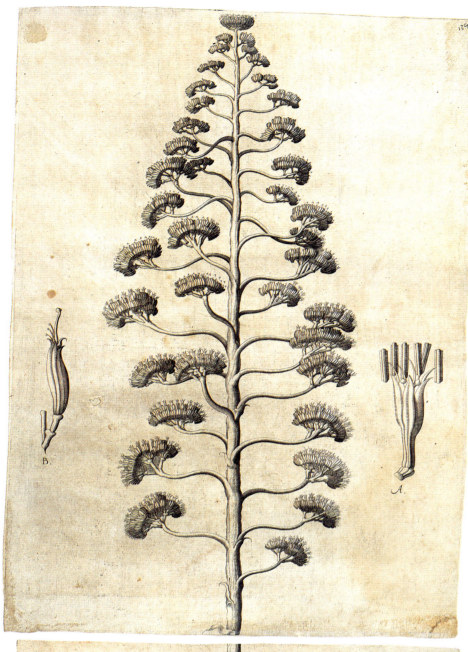

80   A & B DeB, *Agave americana*

AICHSTA

REVERENDISS. ET. ILLVSTRISS.
S. R. I.
cipi. IOANNI. CHRISTOPHORO
Episcopo. Aichstetensi
VRBEM. SVAM. QVAM. IPSE
VIRTVTE. CONSILIO. OPIBVS
ILLVSTRIOREM. REDDIDIT